BEYOND VOM KRIEGE

The Character and Conduct of Modern War

Published by:
University of North Georgia Press
Dahlonega, Georgia

Printing Support by:
Lightning Source Inc.
La Vergne, Tennessee

Cover and book design by Corey Parson.

ISBN: 978-1-940771-71-7

Printed in the United States of America
For more information, please visit: http://ung.edu/university-press
Or e-mail: ungpress@ung.edu

BEYOND VOM KRIEGE

The Character and Conduct of Modern War

R. D. Hooker, Jr.

Foreword by Lieutenant General H. R. McMaster, USA

To Beverly
My shield and strength

War is the father and king of all: some he has made gods,
and some men; some slaves and some free . . .

—Heraclitus

CONTENTS

Foreword ix
Acknowledgments xi
Introduction xiii

1 The Grand Strategy of the United States 3

2 The Strange Voyage: A Short Précis on Strategy 40

3 National Security Decision Making and the NSC 56

4 The Character and Conduct of Modern War 73

5 The Future of Deterrence 92

6 Soldiers of the State: Reconsidering American Civil-Military Relations 109

7 A Critique of the Maritime Strategy 129

8 The Role of Airpower in Modern War 146

9 The Role of Land Power in American National Security 183

10 Presidential Decision Making and Use of Force: Case Study Grenada 201

11 Hard Day's Night: A Retrospective on the American Intervention in Somalia 217

12 Offshore Control: Thinking about Confrontation with China 241

13 Iraq and Afghanistan: Reflections on Lessons Encountered 253

14 NATO in Crisis 272

About the Author and Contributors 284

Foreword

America's military has made the development of leaders its top priority. That is because the complex environments in which U.S. forces must operate and the broad range of potential contingencies for which they must prepare demand courageous, adaptive, and innovative leadership. While the issues that R. D. Hooker, Jr., explores in this volume are intrinsically important, equally important is his example of scholarship and professionalism across four decades of service. The U.S. Army emphasizes "the development of expert knowledge and the ability to use it with the right moral character that sustains excellence in every endeavor, at home and abroad." If young leaders are looking for an example of how a true professional develops expertise over time through self-education—thinking, reading, discussing, and writing about the issues that bear on their responsibilities—R. D. Hooker provides that example in this collection.

The essays in this volume provide a window into an officer's career during a period in which the Cold War ended, U.S. forces operated in context of a broad range of missions during the strategically ambiguous "post-Cold War decade of the 1990s, and mass murder attacks on the United States on September 11, 2001, initiated America's longest war." He writes with unique authority because he served in many of the conflicts to which U.S. forces were deployed in that period and because he combines that experience with academic training and an ability to write clearly about them.

Consistent with Sir Michael Howard's guidance on how military professionals should develop their own theory or understanding of war, Hooker's essays approach the subject from perspectives that consider the subject of war and warfare in width, depth, and context. Howard enjoined military leaders to study war first in width to observe how warfare developed over history. Next, leaders should study armed conflict in depth through the examination of campaigns to reveal the complex causality of events as the "tidy outline dissolves," and we "catch a glimpse of the confusion and horror of real experience." And lastly, to study in context because wars cannot be understood without consideration of their social, cultural, economic, human, moral, political, and psychological dimensions and because "the roots of victory and defeat often have to be sought far from the battlefield."

This is a book of great value to military professionals, civilian policymakers, and all those interested in issues of national and international security. The essays will help deepen understanding of and provide perspective for contemporary issues. And they also serve as an example of what it means to be a true military professional. Admiral James Stavridis, the former Supreme Allied Commander Europe (SACEUR), enjoined officers to "live well, write about it, and write it well." R. D. Hooker has done just that.

H. R. McMaster
Lieutenant General, U.S. Army (Retired)

Acknowledgments

This volume represents an intellectual journey that began at West Point and continued throughout a long military career and beyond. Though a career soldier, the desire to think and write, first awakened at West Point, remained with me through four decades of service. Graduate school at the University of Virginia provided me the intellectual equipment to begin to think seriously about strategy and the use of force in international relations. There, I studied under Kenneth Thompson, S. Neil MacFarlane, Henry Abraham, and many others who helped me build not only a theoretical foundation, but also a practical understanding of international politics and of the juridical and ethical framework which shapes and guides our society and the military's place within it. I followed this experience with a tour in the department of social sciences at West Point, a national treasure whose alumnae include Generals David Petraeus, Brent Scowcroft, John Abizaid, Pete Chiarelli, and many others. On the West Point faculty, I was surrounded by brilliant young officers who would later play pivotal roles in Iraq and Afghanistan.

I wish to acknowledge the mentorship and friendship over many years of the following people, whose advice and support helped shape my thinking, and who directly or indirectly supported this project: Professor Joseph Collins, Professor S. Neil MacFarlane, Brigadier General Mitchell Zais, Lieutenant General H. R. McMaster, Lieutenant General Rick Waddell, Major General Skip Davis, Colonel David Gray, Professor Sir

Hew Strachan, Professor Meghan O'Sullivan, General Mike Scaparrotti, Lieutenant General Doug Lute, Admiral Harry Harris, General David Petraeus, General John Allen, Mr. Robert Bell, Admiral James Stavridis, Brigadier General Mike Meese, Admiral Stansfield Turner, and Professor Harvey Rishikof. I extend grateful appreciation to the editors who have offered kind permission to reprint my work in this forum, to Dr. Jeff Smotherman and the exceptional staff of the National Defense University Press, and to my colleagues at the Institute for National Strategic Studies. I wish also to thank Ms. Brittany Porro, who provided valuable editorial assistance.

These reflections, then, are offered respectfully for those who must fight, and even more for those who must contend with the awful questions of peace and war. For any errors of omission or commission, I am of course entirely responsible.

R. D. Hooker, Jr.
National Defense University
Washington, D.C.

Introduction

America's performance in war since 1945 has been mixed, at best, and this volume attempts to explore both the virtues and the flaws that attend American national security and strategy making. America's advantages are many: the world's leading economy; a strong and innovative technology base and skilled workforce; an unmatched military, particularly in the air and on the ground; an invulnerable nuclear deterrent; a public that is both confident in and supportive of its military institutions; a large pool of qualified young people; a dense network of allies and partners that together account for much of the military capacity on the planet; and a favorable geostrategic position. These attributes propelled the U.S. to dominance in the twentieth century, enabling successful outcomes in both world wars and the Cold War. Yet since 1945, America has often faltered in conflict, its strategic performance falling well short of its promise. Why is this so, and what can be done about it?

I joined the national security enterprise in a period marked by a struggling economy; internal divisions; an under resourced, ill-disciplined, and poorly-managed military; and a loss of national confidence that took a decade to overcome. By the early 1990s, much had changed. A rejuvenated U.S. military had recovered its swagger, scoring successes in Grenada, Panama, and the Gulf War, and presiding over the end of the

Cold War. Scholars wrote of a new era of "unipolarity"[1] while politicians heralded a "New World Order."[2]

The warm afterglow of victory in the Cold War soon faded. After a good start in the waning days of the Bush '41 presidency, the U.S. intervention in Somalia experienced epic failure in the early days of the Clinton administration, recalling Wellington's admonition that "for a great power there are no small wars." Swinging from overconfidence to extreme caution, Clinton pursued a hands off policy towards the Balkan Wars and to the Rwandan genocide. In the midst of a dramatic downsizing of the U.S. military, the U.S. found itself engaged in large scale and enduring "stabilization" deployments in Bosnia and later Kosovo, placing surprising stress on a much reduced U.S. Army. In Bosnia, stabilization efforts were successful in enforcing the Dayton Accords and preventing open conflict, but the resulting political settlement has proven unstable and untenable. In Kosovo, Serbian military forces were evicted after a short, intense air campaign—but more Kosovar civilians were killed after the start of military operations than before, while western-supported Kosovo independence has resulted in a weak and fragile Kosovo and an intransigent Serbia, backed by its traditional "big brother," a resentful Russian Federation.

Far from ushering in an era of calm, the post-Cold War period and the breakdown of bi-polarity brought with it an explosion of new, weak states and non-state actors. While the cataclysmic great power clashes of the twentieth century receded, brushfire wars and terrorism exploded. I was working in the Pentagon on 9/11, which shattered an American sense of security in the homeland, ushering in a generation of counter-terrorism and counter-insurgency in Iraq, Afghanistan, Syria, Africa, and the Philippines that dangerously over-stretched the U.S. military and imposed huge financial and human costs.[3] Neither Iraq nor Afghanistan

1 "Since the collapse of the Soviet Union, the United States enjoys unparalleled military power. The international system is therefore unipolar." See Nuno P. Monteiro, *Theory of unipolar politics*, Cambridge Studies in International Relations, Yale University, 2014.

2 See President George H. W. Bush, *Address before a joint session of the Congress*, September 11, 1990.

3 R. D. Hooker, Jr.. and Joseph Collins, 2016, *Lessons encountered: Learning from the Long War*, pp. 421–440, Washington, D.C.: National Defense University Press.

led to clear gains—while Saddam Hussein was eliminated and the terrorist safe haven in Afghanistan removed, Iraq is today dangerously unstable and subject to Iranian influence, while the Taliban remains resilient in the face of weak Afghan governance. Once again stepping back from more than a decade of wrenching combat operations, the U.S. watched from the sidelines as the promise of the Arab spring degenerated into the chaos of the Syrian civil war (with a half million civilian dead and more than thirteen million refugees and displaced persons). Today, despite American economic and military dominance, the U.S. finds itself challenged by revanchist powers China and Russia as well as rogue states like North Korea and Iran. Importantly, we face pressing challenges in the cyber and information domains and, more generally, in the "gray zone" that put important U.S. interests at risk.[4] Despite far greater economic resources, we face real threats from these states as well as from transnational criminal organizations and proliferating terrorist groups. Our global dominance and many allies have not created a stable and ordered international system.

How is it that our economic and military power so often fails to translate into success in war? We can begin with how America approaches strategy. In broad terms, there is great continuity in American "grand" strategy. Great military strength, the largest economy in the world, strong alliances and partnerships, forward basing, a powerful and innovative industrial and technology base, and an invulnerable strategic nuclear deterrent underlay the great success of the Cold War and ensured America's global preponderance. But in many specific cases—Korea, Vietnam, Somalia, Iraq, and Afghanistan—"victory" proved elusive. Why?

The American military's ability to attack and destroy targets—to "kill people and break things"—is clearly not the problem. In almost every example on record in recent decades, the U.S. has prevailed in battles and engagements, even when outnumbered and outgunned. Rather,

4 Gray zone challenges have been described as "attempts to achieve one's security objectives without resort to direct and sizable use of force." See John Schaus et al., July 2018, "What works: Countering gray zone coercion," *Center for Strategic and International Studies*.

our ability to see, understand, and define the strategic challenge is all too often flawed. For example, in the Korean conflict, U.S. political and military leaders failed to discern—despite many warning signs—that China would not permit North Korea to be defeated and occupied, or that our massive air and seapower might not translate into success on land. The means we were prepared to bring to bear to cope with China's intervention were manifestly inadequate to achieve the desired end. In Vietnam, Iraq, and Afghanistan, we failed to discern that the problems of sanctuary in adjoining countries and incapable and corrupt host nation governments—themselves drivers of the conflict—could not be solved with the means we prepared. Too often, our "ends" proved aspirational and unrealistic, our "means" well below the level required. If there is a lesson to be found in these three conflicts, it is that America is poorly suited to large scale counter-insurgency campaigns, which almost by definition rarely engage truly vital U.S. interests.

In this regard, there is clear evidence in our post-war history of a recurring political dynamic which has hindered American success in war. In each of our major military interventions since 1945, a singular event sparked (or was thought to have sparked) an urgent need for military action.[5] With the single exception of the Gulf War, resulting military operations did not lead to clear success. Instead, mounting costs and inconclusive results (as well as other competing strategic priorities) led successive administrations to make decisions that staved off defeat without enabling victory. An apparent inability to comprehend the strategic challenge accurately, and to link means to desired ends in concrete ways, prevented ultimate success that crippled and even destroyed presidencies. While intensely polarized politics is partly to blame, this tendency towards "no win, no lose" approaches represents a clear trend in American

5 In Korea, the North Korean invasion in June 1950; in Vietnam, the collapse of the U.S.-led advisory effort following the Diem coup; in the Gulf War, Saddam's invasion of Kuwait; in Afghanistan, the 9/11 attacks. The Bush '43 administration attempted to link the invasion of Iraq in 2003 with 9/11, an assertion that failed to gain traction and was subsequently replaced with Saddam's supposed possession of weapons of mass destruction.

strategic performance since World War II (WWII). The consequences for the United States have been doleful.

While it is unfair to hold the U.S. military accountable for poor political decisions, it must share responsibility for outcomes. Since the passage of the Goldwater-Nichols reforms in the late 1980s, we have seen much triumphalism about improved "jointness." Yet the reality is quite different. In peacetime, the military services rarely train with each other. Service approaches to warfighting and roles and missions remain grounded in the definitive experiences of WWII, updated with new technology. Even on the battlefield, the services fight hard to preserve their freedom of action relative to each other. Joint doctrine at best papers over sharp disagreements between services, above all with respect to the use of airpower. Lacking Title 10 legal authorities, the Chairman of the Joint Chiefs of Staff (CJCS) can cajole and suggest but cannot demand material changes in service culture, while strong congressional influence and typically short tenures and high turnover limit the ability of the secretary of defense to address the problem.[6]

A few examples are illustrative. A close look at the department of the Navy reveals by far the largest and strongest fleet in the world but also (in the form of the Marine Corps) a land force larger and more capable than all but a handful of the armies on the planet.[7,8] The Navy's air arm is again larger and more capable than most of the world's air forces—while Marine aviation by itself can make the same claim. The possession of powerful air, land, and sea forces within a single military department translates into a high degree of autonomy which the Navy jealousy guards. Soaked in the

6 When interviewed by the author for a study of the Iraq and Afghanistan campaigns, then-CJCS General Martin Dempsey noted that during his four-year term, he served three different Secretaries of Defense (Leon Panetta, Chuck Hagel and Ashton Carter).

7 The U.S. Marine Corps alone, with 184,000 troops, is larger than the entire armed forces of Germany (178,000), France (161,000) or the U.K. (150,000). Its aviation component boasts twice as many combat aircraft as found in the air forces of any of these three. *The Military Balance 2018*, International Institute of Strategic Studies.

8 President Truman famously called the Marine Corps "the navy's little army that talks navy." In fact, it is one of the most versatile and powerful military institutions in the world.

tradition of Mahan, the sea services hold to a vison of victory through seapower that remains very much alive and well.

Similarly, the U.S. Air Force, bathed in the theories of Douhet and Mitchell, sees itself as fully capable of achieving decisive "strategic" outcomes independently from the other services. Aided by procurement budgets far greater than the Army, and free to operate with great autonomy, the Air Force—like the Navy—can deploy and conduct operations in a theater of war largely independently. In both Iraq and Afghanistan, air force and marine units were not tasked organized under the theater commander, but instead reported to the combatant commander in Tampa.[9] The same was true of special operations forces for most of both conflicts. As a result, theater joint force commanders in both Iraq and Afghanistan found themselves in the unenviable position of not "owning" the air force, marine, and special operations forces operating in their battlespace. In such circumstances, unity of effort proved difficult to achieve, while unity of command was altogether absent.

Such behavior aligns comfortably with organizational theory, which holds that organizations strive for freedom of action and for the greatest possible share of resources. There is no mystery here. But the drive towards autonomy must collide with the requirements of effective strategy, which seeks the most efficient and effective use of available resources to accomplish demanding and complex tasks. This strategic disability recurs in all American conflicts. The national interest must trump service parochialism. Too often, it doesn't.

Alone among America's military services, the U.S. Army can be exempted from this general critique. Unlike the Navy and Air Force, the Army is uniquely dependent on its sister services. It cannot move itself to the theater of war. It cannot defend against strong enemy air forces. It requires secure sea lanes of communication to survive. Even the most junior army officer exists in a milieu where external help in the form of

9 *Lessons encountered*, p. 10.

artillery, aviation, logistics, intelligence and much else—all found outside the basic infantry or tank unit—may literally spell life and death. From birth, the army officer is bred to be anything but autonomous, and this fact defines the Army's culture.[10] To insist on service autonomy in the theater of war does not occur to Army commanders, who have no professional experience of it.[11]

Perhaps as a result, the Army finds itself a consistent loser in the budget and acquisition battles that largely define success "inside the beltway." By any measure, America is strongly preponderant in the air and at sea. On land, the picture is rather different. While well-equipped and well trained, the U.S. Army fields legacy systems that are now some four decades old. Its armor and artillery communities suffered deep cuts as the Army reorganized following 9/11; today the Army is predominantly light infantry, with far less striking power than formerly. In size, the Army has similar liabilities. Manned at almost 800,000 soldiers and 18 divisions in the 1980s (half of which were armored or mechanized), the Army today fields less than 500,000. Though tasked with global responsibilities, it cannot realistically fight more than a single "major regional contingency" at a time. And despite claims to the contrary, air and seapower cannot supply this deficiency in large scale campaigns fought on land, as seen in Korea, Vietnam, Iraq, and Afghanistan. In several essays, this volume emphasizes this theme. America's military is out of balance, and we have paid a price accordingly.

In the post-war period, there is one striking example of strategic

10 These aspects of service culture are described at length in Carl Builder's *The masks of war: American military styles in strategy and analysis* (Baltimore: The Johns Hopkins University Press, 1989).

11 In the Gulf War, Army and Marine Forces fought separately and not under Land Component Command, with the 1st (UK) Armored Division interposed between them. See P. Mason Carpenter, *Joint Operations in the Gulf War: An Allison Analysis,* Air University School of Advanced Airpower Studies, June 1994. In the invasion of Iraq in 2003, the Army fought west of the Euphrates River, the Marines east of the river. In both Iraq and Afghanistan, Air Force units were task organized under U.S. Central Command's Air Component Commander, a 3-star based in Qatar and reporting to Tampa, not the theater commanders in Baghdad and Kabul. These examples illustrate how service rivalries are finessed in wartime at the expense of unity of command.

success in major theater war: Operation Desert Storm in 1991. The Gulf War was marked by clear, limited political objectives ("eject Saddam from Kuwait"); overwhelming force; strong support from the Congress, the public, and allies; sound and intelligent planning from the national to the tactical level; and extraordinarily competent execution. Casualties were extremely low, while the campaign was won in a matter of weeks—the ground phase in only four days. What was different here?

In a word, the answer must be "leadership." President George H. W. Bush came to office as arguably the most experienced and qualified commander-in-chief in modern American history. Brent Scowcroft, his national security adviser, has been described as "the gold standard" in this critical position. General Colin Powell, the Chairman of the Joint Chiefs—himself a former national security adviser—is generally considered the most outstanding chairman ever. General Norman Schwarzkopf, the commander of U.S. Central Command, provided driving and intelligent command and control, suppressing service rivalries and proving himself a master of joint and coalition warfare. In many ways, Desert Storm represents a blueprint for success. Regrettably, its lessons have been largely ignored by later administrations.

A strategic education and experience in managing wars are not normally found on presidential resumes, a phenomenon compounded by the American custom of salting government departments and agencies three- and four-levels deep with political appointees with varying degrees of expertise. This political reality has consequences. In surveying the history of America at war, one is struck by a strange sense that we must learn and relearn the same lessons over and over again. As Sir Hew Strachan and others have pointed out, "strategy is a profoundly pragmatic business."[12] In its essence, it need not be diabolically difficult. Yet war severely punishes fundamental mistakes—the inability to identify the problem, poor assumptions, failure to link means with ends,

12 Sir Hew Strachan, 2014, *The Direction of War: Contemporary Strategy in Historical Perspective*, p. 12, Cambridge: Cambridge University Press.

failure to learn and adapt. Our often ahistorical approach is abetted by an apparent inability to see the problem from the adversary's point of view. This problem of "filters"—the tendency to assume that one's opponent and one's allies see the world as we do—is a besetting sin in American strategy making.

In addition to the major campaigns mentioned above, the U.S. has engaged in many smaller ones in the post-WWII era, again with varying degrees of success. Military interventions in Lebanon in 1958 and the Dominican Republic in 1965 were judged to be generally successful. The Mayaguez incident in 1975, the Iran rescue mission in 1980, and the Marine intervention in Lebanon in 1983 must be classed as failures. The invasion of Grenada in late 1983 accomplished its objectives but revealed serious problems with joint operations and a high number of friendly fire casualties. The invasion of Panama in 1989 showed the U.S. military and the Bush-Scowcroft team at its best, as a challenging and complex operational plan was carried out in short order and with minimal casualties. A later intervention in Somalia in 1993 ended in disaster, despite an auspicious beginning, while "peace enforcement" operations in Bosnia in 1995 and the Kosovo air campaign (and subsequent stability operations) in 1999 accomplished their objectives at low cost. Below the threshold of major combat operations, the U.S. record since 1945 is therefore mixed, and understanding these outcomes requires a more detailed understanding of the particulars. In general, however, overwhelming force can be seen to have a quality all its own, "smothering" friction, intimidating opponents and enabling successful outcomes much more often than not. Epic failures like Desert One and Mogadishu illustrate the dangers inherent in complex operations involving multiple services, far from the homeland and involving only small, light forces.

After a generation of counterterrorism and counterinsurgency, the United States is once again focused on deterring state-on-state conflict and major theater war.[13] China, Russia, Iran, and North Korea are the focus of these efforts, as explicitly laid out in current National Security and National Defense Strategies. Big military budgets are back, and the U.S. is investing in some of the most advanced, and most expensive, technologies available. Yet troubling concerns remain.

First, the U.S. national security enterprise shows limited enthusiasm for recognizing and solving service rivalries, whose deep roots and persistence continue to hinder effective strategy and warfighting. Next, the costs of acquiring new systems and technologies have exploded, along with personnel costs and a massive expansion of headquarters and defense agencies since 1945 that sap the fighting strength of the U.S. military.[14] Our inability to deal decisively with low tech opponents like the North Vietnamese, al-Qaeda, and the Taliban suggest that smaller, more exquisite and more expensive forces combined with more and more command and control may not be the answer. A related but critical worry is that domestic entitlement spending, which dwarfs defense spending, will begin to crowd out all other government spending within a generation unless checked—and there are no signs so far that either political party is willing to take this on.[15]

As for better strategy making, let us begin with a better understanding of war. In America, we resort to war too often, win too infrequently, and comprehend war too poorly. The first lesson is that war is a poor vehicle for solving inherently political problems. War done right can serve the ends of policy, by helping to set conditions for successful political outcomes. But it cannot substitute for political solutions like better governance, rule of law, or fair elections. Too often, we have tied our military and political fortunes to corrupt and failing regimes. A better approach, perhaps, is to

13 See *The National Security Strategy*, 2017, Washington, D.C.: The White House, and *The National Defense Strategy*, 2018, Washington, D.C.: Department of Defense.

14 For a detailed assessment of these issues, see the author's *Charting a course: Strategic choices for a new administration*, pp. 61-82, 2016, Washington, D.C.: National Defense University Press.

15 See Michael Meese, *Strategy and Force Planning in a Time of Austerity*, May 2014, Washington, D.C.: National Defense University Press.

fight less often, for clearer objectives, with stronger forces and stronger support from our voters and allies. Presidents will always be tempted to reach for the sword or the button as an immediate answer to an urgent problem, in recent decades unencumbered by congressional or judicial checks. Yet war has its own nature and will get out of hand if permitted. For the soldier and the president alike, war is about survival—and the struggle for survival is impatient of limits. Here Churchill's admonitory description of war as a "strange voyage" should be heeded.

Successful strategy, therefore, begins by understanding the nature of the conflict.[16] At the outset, we must carefully define the problem, consulting the important national interests that may be engaged, and resisting the impulse to set aspirational, vague goals or to resort to force when other approaches may suffice. Before we rush to generating courses of action, we must gather the facts, and, where facts are missing, make sound assumptions about capabilities, intentions, and risks. We must strive to view the case from the perspective of our adversaries if we are to have any hope of understanding their actions and reactions. We must link means to ends, and where the available means fall short, we must adjust our ends or increase our means. We must devise metrics— measures of effectiveness—so we can judge our progress and adjust if necessary. At all times, we must weigh the support of our voting publics, of our legislatures, of our friends and allies, and of international public opinion. Finally, if we decide on war, we must wage it with a determination to win and, if at all possible, to win quickly and decisively. Fail at any one of these steps and overall failure is probable.

These, then, are the key points of the essays collected in this volume. They are offered not with certainty but with a measure of humility, as

16 "The first, the supreme, the most far reaching act of judgment that the statesman and commander have to make is to establish . . . the kind of war on which they are embarking, neither mistaking it for, nor trying to turn it into, something that is alien to its nature." "No one starts a war—or rather, no one in his senses ought to do so—without first being clear in his own mind what he intends to achieve by that war and how he intends to conduct it." Carl von Clausewitz, (1976). *On War*, pp. 88, 579. Michael Howard and Peter Paret (Eds.). Princeton, NJ: Princeton University Press.

painful lessons painfully learned. "War," Heraclitus reminds us, "is the father and king of all." Would that it were not so. Yet as Lincoln advised

> Human nature will not change. In any future great national trial, compared with the men of this, we shall have as weak and as strong, as silly and as wise, as bad and as good. Let us therefore study the incidents of [war] as philosophy to learn from and none of them as wrongs to be revenged.[17]

17 Speech by President Abraham Lincoln, November 10, 1864.

Part I

1 The Grand Strategy of the United States

Does the United States have a grand strategy? Many scholars are doubtful, while others debate its definitions, applicability, or efficacy when applied to specific cases. Here the author argues that American grand strategy has deep roots and enduring significance over time, though not all strategists understand or apply its prescriptions.[1]

From the earliest days of the Republic, the outlines of an evolving American grand strategy have been evident in our foreign and domestic policy.[2] Much of that history continues to inform our strategic conduct, and American grand strategy therefore rests today on traditional foundations. Despite a welter of theory and debate, grand strategy as a practical matter is remarkably consistent from decade to decade, its means altering as technology advances and institutions evolve but its

1 This essay originally appeared as a National Defense University Institute for National Strategic Studies *Strategic Monograph* in Fall 2014.

2 Defining "grand strategy" is admittedly onerous. Colin Gray defines it as the "purposeful employment of all instruments of power available to a security community." Robert J. Art excludes non-military instruments from grand strategy, while Christopher Layne says simply "the process by which the state matches ends and means in the pursuit of security." Sir Hew Strachan sees grand strategy as forward looking, aspirational, and oriented on preventing or managing great power decline. Edward Luttwak is particularly opaque: "Grand strategy may be seen as a confluence of the military interactions that flow up and down level by level . . . with the varied external relations that form strategy's horizontal dimension at its highest level." See Gray, *War, peace and international relations—An introduction to strategic history* (Abingdon and New York: Routledge), 2007, p. 283; Art, "A defensible defense," *International Security,* Spring 1991, 7; Layne, "Rethinking American Grand strategy: Hegemony or balance of power in the 21st century," *World Policy Journal,* November 1998, p. 8; Strachan, "Strategy and contingency," *International Affairs, 87*(6), 2011, pp. 1281–1296; and Luttwak, Strategy (Cambridge, MA: Harvard University Press), p. 179.

ends and ways showing marked continuity.

Grand strategy can be understood simply as the use of power to secure the state.[3] Thus it exists at a level above particular strategies intended to secure particular ends, and above the use of military power alone to achieve political objectives. One way to comprehend grand strategy is to look for long-term state behavior as defined by enduring, core security interests and how the state secures and advances them over time. In a way, this means that what the state does matters more than what the state says. Grand strategy is therefore related to, but not synonymous with, National Security Strategies, National Military Strategies, Quadrennial Defense Reviews, or Defense Strategic Guidance. Grand strategy transcends the security pronouncements of political parties or individual administrations. Viewed in this light, American grand strategy shows great persistence over time, orienting on those things deemed most important—those interests for which virtually any administration will spend, legislate, threaten, or fight to defend.

The Roots of American Grand Strategy

American grand strategy cannot be understood without historical grounding. Prior to the Revolution, the defense of the American colonies was left to the British Crown and, for local defense, to the colonial militia. Contention between the great powers (Spain, the Netherlands, France, and Great Britain) on the North American continent bred an enduring distaste among the colonists for international intervention in the western hemisphere. Pre-Revolutionary warfare was endemic and near constant in North America, on the one hand fostering a familiarity with conflict, but on the other, a distrust of standing forces that would condition American strategic thought for several centuries.[4]

3 "Strategy" is more properly limited to "the deployment and use of armed forces to attain a given political objective." See Michael Howard, "The Forgotten Dimensions of Strategy," *Foreign Affairs*, Summer 1979, p. 975.

4 The list includes the Anglo-Dutch Wars (1665–1667, 1672–1672, 1780–1784), King William's War (1688–1697), Queen Anne's War (1702–1713), the War of Jenkin's Ear (1739–1748), King George's War (1744–1748), and the French and Indian War (1754–1763). The first clash in North

As the United States became more firmly established, this impulse found expression in the Monroe Doctrine and in a general aversion to involvement in European wars that dated from Washington's first administration.[5] This partly stemmed from military and economic weakness, but the desire not to become enmeshed in the politics of great power rivalry also played a key role. America was fortunate not to be drawn more deeply into the French Revolutionary and Napoleonic Wars, and thereafter the desire to pursue continental expansion and to exclude further European colonization of the hemisphere shaped our policy and strategy for the rest of the nineteenth century.[6]

American grand strategy also carried a defining ideological component from the beginning. While generally pragmatic, early American political and military leaders were strongly influenced by the ideals of the Enlightenment and the Revolution and by an emerging American political consciousness.[7] Since the Revolution, most American conflicts have been articulated and justified with some reference to this founding ideology, lending a distinctive, normative dimension to American strategy and strategic culture. Sometimes described as "American Exceptionalism," this component has been seen by some as an impulse to promote democratic values and the rule of law abroad as well as at home. By others, it's seen as an excuse for intervention.[8]

America between European powers was the 1565 Spanish massacre of French Huguenots at Fort Caroline in Florida.

5 The War of 1812 entangled the United States peripherally in the Napoleonic Wars over questions of trade restrictions with France, impressment of U.S. sailors (many of whom were British born but naturalized American citizens) on the high seas, and British support for Indian tribes resisting expansion into the Northwest territories. Expansion into Crown territories in Canada was also a war aim. Though arguably a victory, the War of 1812, which saw the burning of Washington, D.C., and numerous other defeats, confirmed the view that military engagement with the great European powers was not in American interests.

6 Eugene V. Rostow, *A breakfast for Bonaparte: U.S. national security interests from the heights of Abraham to the Nuclear Age* (Washington, D.C.: National Defense University Press, 1993), pp. 143–143.

7 Ibid, p. 78.

8 Described by Lipset as a new American ideology, based on notions of personal liberty, egalitarianism, individualism, republicanism, populism, and laissez-faire. Seymour Martin Lipset, *American exceptionalism: A double edged sword,* (NY: Norton, 1997), pp. 17–19.

Although our historical narrative emphasizes reliance on local militia forces, as early as the Revolution regular forces, or volunteer units raised outside the militia organizational structure, have formed the center of gravity of America's military establishment as far back as the Revolutionary War.[9] At least through the Korean conflict, for all significant campaigns the pattern or cycle of America at war featured small regular forces, an expansion of the army through a combination of militia call-ups, volunteering and conscription during the conflict, and then a drawdown or return to pre-war levels. This original aversion to large standing forces was undoubtedly rooted in the English Civil War; many of the original settlers in the American colonies came to the New World to escape the repression and incessant conflict they found in the Old, and those memories became firmly imprinted in their cultural DNA.

Through the nineteenth century, the United States grew and evolved as a rising regional power, only achieving great power status at the beginning of the twentieth century. The collapse of the Spanish empire in South America and the 1867 emergence of Canada as an independent commonwealth nation accelerated an effective end to European presence in the Western Hemisphere that was rendered final with the ejection of Spain from Cuba and Puerto Rico in 1898.[10] Territorial expansion through the Louisiana Purchase, the Mexican-American War, the Alaska Purchase, and the Indian Wars completed the process of continental growth, accompanied by large scale immigration from Europe, the transcontinental railroad, a growing and powerful mercantile capacity and industrialization on a broad scale. Thus, the conditions were set for America's evolution into a superpower in the following century.

Overshadowing everything else in the nineteenth century is the American Civil War. Vast in scope and scale, the Civil War fundamentally

9 For example, the Second Seminole War (1836) was fought with 10,000 regulars and 30,000 volunteers. Local militia forces played no significant role. Russell F. Weigley, *The American way of war* (Bloomington: Indiana University Press, 1973), p. 68.

10 The Monroe Doctrine (1823) had specifically excluded Cuba.

challenged the survival of the nation and its constitutional system. More Americans died in the Civil War than in all other U.S. wars. Over the course of the conflict, very large land and naval forces were raised, conscription was invoked, and modern technologies like mass production; military railroads; the telegraph; breech-loading, rifled artillery; repeating rifles; and iron-clad warships were introduced. Modern military professionalism and generalship replaced the notion of the talented amateur. Profound political questions were settled, most importantly the central role and importance of the federal government and the president as Chief Executive and Commander-in-Chief. There would be no going back.

Though the military establishment returned to pre-war levels following the Civil War, the precedent of mass mobilization under an organized war and navy department and professional generals and admirals had been well established. Professional military education took root, notably at the Naval War College at Newport, Rhode Island and at the Army's School of Application for Infantry and Cavalry (later the Command and General Staff College).[11] Up to the Spanish American War, the Army performed essentially constabulary duties, while the Navy steadily evolved towards a modern, capable, technically proficient arm of the service with a coherent doctrine. By the end of the nineteenth century, the general tenets of American grand strategy were well established and consistently applied by presidents and congressional leaders of both parties. The overriding principle was, and remains, the protection of American territory, citizens, constitutional system of government, and economic wellbeing. These "vital interests" were secured and enabled in the 1800s through protection of trade and freedom of navigation on the oceans; a prohibition against European military intervention in the Western Hemisphere; a capable navy; a small but professional army, capable of rapid expansion in time of crisis; and a readiness to provide

11 Weigley credits Newport with being "the first institution of its type anywhere in the world." *The American way of war*, p. 172.

support to civil authorities when needed. Protected by two vast oceans, with an industrialized and increasingly global economy and a large and growing population (enabling a potentially huge land force if threatened), the United States generally enjoyed a stable security environment.

A Century Like No Other

The new century would transform American grand strategy in different but comparable ways. By a wide margin, the twentieth century would prove to be the most catastrophic in history. The Spanish American War, while revealing many shortcomings in organization and supply for the land forces, showcased a powerful and competent navy with global reach. It also made the United States an imperial power with newly-won possessions in the Caribbean (Puerto Rico) and the Pacific (the Philippines and Guam). America had now moved decisively onto the world stage.

In the following decade, it became clear that war loomed in Europe as armies assumed massive proportions, professional general staffs perfected the machinery of mobilization, and industrialization and advancing technology equipped armies and navies for large scale, protracted war. The United States, preoccupied with colonial concerns in the Philippines and protected by an impressive fleet and oceans on either side, genuinely pursued a neutrality that would eventually founder on two key strategic dilemmas: (1) the protection of trade and markets, and (2) the potential rise of a hostile power in control of the European landmass. American pride was certainly touched by unrestricted submarine warfare, but what could not be borne was the isolation of U.S. commerce from European markets, or the prospect of German control of all of Europe's economic and demographic resources. At that point, Germany could conceivably threaten the continental U.S., both militarily and by setting the terms of trade. While cultural and ideological affinities with European democracies played important roles, and a politically powerful isolationist movement offered resistance, these life-and-death strategic considerations compelled America's entry as an active belligerent.[12]

Unlike WWII, America was no "arsenal of democracy" in WWI. Once committed to war, U.S. grand strategy stressed speed over mobilization of the industrial base and a deliberate buildup of troops and material. Getting large field forces to France in time to prevent an Allied collapse was the driving strategic imperative. France, the United Kingdom (U.K.), Italy, and Russia supplied their own weapons and equipment. American forces were largely equipped by our allies, except for small arms. Still, the introduction of a one million-man U.S. field army, just as Germany's defeat of Russia enabled the transfer of huge forces to the Western Front, proved decisive. In only three months of large-scale combat, the U.S. suffered heavy casualties, but the arrival of the Americans proved decisive to victory. By war's end, the United States had moved to the fore as a great power and a guarantor of the international order.[13]

The Armistice was followed in the 1920s by massive demobilization and in the 1930s by economic collapse, repeating the familiar pattern of putting the Army in caretaker or cadre status. In contrast, though limited by treaty restrictions, the Navy pursued the development of carrier aviation and long-range submarines, while inside the Army Air Forces, the foundations of a strategic bomber force were laid. A resurgent Germany, well ahead of its rivals with newly developed armored formations and a modern air force, once again raised the specter of a non-democratic power possessing the European continent and directly threatening the continental United States. This time, however, the strategic challenge was far more complex and dangerous. In Asia, a modern and bellicose Japanese Empire invaded China and looked ready to challenge American economic and territorial interests in the Pacific, while an ideologically virulent Soviet Union raised huge forces even as it savagely repressed

12 Rostow, p. 245.

13 WWI casualties included 37 million civilian and military dead and wounded, including 120,000 U.S. dead and 205,000 wounded.

millions of its citizens, killing more than 14 million peasants in the forced collectivization of the 1930s. At the outbreak of war in 1939, America found itself once again with a small and unprepared land force and with unready allies.

U.S. grand strategy in WWII aimed at the defeat and destruction of Germany and Japan, not as ends in themselves but as a necessity to the reestablishment of a stable international order, a prosperous global economic system, and a U.S. population that would be free from military threat at home and abroad.[14] This necessitated strong support for allies—even unsavory ones like the U.S.S.R. , which proved essential to victory—massive mobilization, and an economic and industrial effort unparalleled in world history. Even in retrospect, the U.S. effort beggars belief. By war's end, the U.S. Navy was larger than the combined fleets of every other combatant nation, owning more than 70% of the naval strength in the world. The U.S. Army, ranked seventeenth in size in 1939, grew to more than eight million soldiers and ninety combat divisions. The Army Air Forces boasted 80,000 aircraft. American ships, planes, and tanks were among the most reliable and effective in the world and were supported by a supply system unrivalled on the planet. Beginning slowly, the U.S. and its allies advanced progressively throughout the war, gaining the initiative in the Pacific in 1942 and in Europe in 1944.

U.S. grand strategy, as distinct from theater strategies in Europe and Asia, focused first on keeping the British, Russians, and Chinese in the war while the American buildup gathered momentum.[15] Success was far from assured. In 1940, following an embarrassingly inept Allied performance in Norway, France fell and the British were soundly defeated, narrowly

14 President Roosevelt put it succinctly in a "fireside chat" early in the war: "If Great Britain goes down, the Axis powers will control the continents of Europe [and] Asia, and the high seas—and they will be in a position to bring enormous military and naval resources against this hemisphere." Russell E. Buhite and David W. Levy (Eds.), *FDR's fireside chats* (Norman: University of Oklahoma Press, 1992), p. 163.

15 This approach was indisputably successful. For every U.S. soldier killed in the war, the Germans lost fifteen, the Russians fifty-three. John Lewis Gaddes, *Strategies of containment: A critical appraisal of postwar American national security policy* (London: Oxford University Press, 1982), p. 8.

escaping annihilation. 1941 saw further humiliations in Greece, Crete, and North Africa, while Russian forces were driven back to the gates of Moscow with millions killed, wounded, and captured. In 1942, Singapore surrendered, the largest capitulation in British history, followed by the near destruction of the U.S. Pacific Fleet at Pearl Harbor. Looking back, Allied victory seems inevitable. At the time, it was anything but.

Over time, enemy strategic missteps, the accumulation of experience at all levels, and, most tellingly, the sheer size and mass of Allied forces (particularly Russian and American) began to tell. It is difficult to argue that man for man and unit for unit, the Allies eventually became better than our adversaries (at least in Europe). [16] What is incontestable is that American mass in all domains proved decisive. Coalition warfare on a global scale, enabled by the most powerful economy and industrial base in history, proved a war-winning combination.

Any sound analysis of WWII must conclude that, in the end, U.S. material superiority proved the decisive factor.[17] America's sheer ability to produce and transport vehicles, ammunition, food, supplies, and fuel kept its key allies on their feet. U.S. industry produced more than 370,000 planes, more than 100,000 tanks and armored vehicles, and more than 7,000 warships during the war. The ability to mobilize and organize the U.S. economy for global war and to field trained and very strong forces in all domains (sea, air, and land) arguably counted for more than where and how they were used.

American grand strategy in WWII was simple, consistent, and effective. Comprehensive defeat of the enemy was envisioned from the start, with Europe as the first priority. Building up our war capacity at speed, while sustaining critical allies (which forced hard resource choices,

16 The German army retained a qualitative edge right up to the end of the war. In the summer and fall of 1944, U.S. infantry regiments were averaging 100% casualties every ninety days. John English, *On infantry* (New York: Praeger, 1981), p. 79.

17 "The war was decided by the weight of armaments production." Alan Milward, *War, economy and society 1939–1945* (Los Angeles: University of California Press, 1979), p. 75.

especially early on) constituted the focus of effort.[18] As the U.S. built strength, President Roosevelt ruled against dramatic and overly risky suggestions to reinforce General MacArthur in the Philippines in 1942 and later, to attempt a cross-channel invasion of Europe in 1943. Instead, the U.S. patiently set the conditions for strategic success. In the Atlantic, this meant defeating the submarine threat. In Europe, this meant strategic bombing on the largest scale to attack German morale, war production and lines of communication, while preparing and then executing the invasion of the continent. In the Pacific, it meant establishing airfields and naval bases and advancing deliberately across the Pacific in a coordinated campaign to engage and destroy the Imperial Japanese Fleet and commercial shipping preparatory to invasion of the home islands. Overwhelming, Allied strength on the ground, in the air, and at sea forced the collapse of Germany and would have done the same to Japan had the advent of nuclear weapons not terminated the conflict.[19]

At war's end, the United States stood alone as leader of the victorious coalition, the greatest economic and military power in the world. In the immediate post-war period, U.S. advantages were absolute. A booming economy, a formidable strategic air force and navy, and sole possession of nuclear weapons ensured American supremacy, fitting it uniquely for a role as sole superpower. American grand strategy at mid-century continued to rest on the foundations described above and could be summarized concisely as monitoring and enforcing a stable international order and economic system that preserved American sovereignty, security, and prosperity; ensuring the security of the homeland through

18 American material support to allies, who did the brunt of the fighting and suffered far more casualties, was unquestionably the strategic center of gravity of the war effort. Some 75% of all German casualties in WWII occurred on the Eastern Front, while the Chinese army inflicted 2.1 million casualties on Japan, compared to 600,000 by the U.S. The U.S. supplied a staggering 11,450 planes, 7,172 tanks, and 433,000 trucks to the Soviets during the war, as well as armor plating for another 20,000 tanks. Alan Gropman (Ed.), *The big L: American logistics in World War II* (Washington: National Defense University Press, 1997), p. 287.

19 Worldwide, the cost of the war far exceeded WWI, with an estimated 85 million dead from all causes. U.S. deaths totaled 418,000—far below every other major combatant.

nuclear deterrence, alliances, forward deployed ground forces, and air and seapower; and preventing the rise of peer competitors that might challenge its economic and military superiority.[20] The isolationism that had existed as a strain in American foreign policy since the beginning would not disappear altogether, but it would never again contend for primacy in grand strategy.

America's supreme effort in WWII did not lead to peace, and unchallenged American dominance proved transitory.[21] As the U.S. demobilized its army, the Soviet Union maintained a powerful and dangerous military establishment that soon gained a nuclear component that could reach U.S. targets. Despite incredible losses during the war, the U.S.S.R. pursued a ruthlessly disciplined political and military program that soon brought all of Eastern Europe under its sway.[22] In Asia, Communist China finally completed their long civil war, driving the nationalists to Taiwan and solidifying their status as a regional power. Both China and the Soviet Union espoused political doctrines and ideologies profoundly at odds with the values and interests of the West. The stage was thus set for decades of confrontation.

In June of 1950, the United States stumbled into an unexpected confrontation with the Communist bloc when the North Korean army invaded South Korea and took Seoul. Unaccountably, North Korea and its Chinese partners seemed not to fear America's nuclear arsenal. At the outset, the lack of strategic warning, poor military preparedness, and uncertainty over U.S. strategic aims muddled the American response, contributing to the indecisive outcome. Although still in possession of

20 This meant rebuilding Germany and Japan as allies and economic partners, but also restraining their military power—in effect containing them as well as the U.S.S.R. Layne, "Rethinking American grand strategy," p. 12.

21 According to Dean Rusk, "by the summer of 1946 . . . we did not have one combat ready division or air wing in the US military." Cited in Rostow, p. 355.

22 "The Red Army suffered 29,629,205 casualties from June 22, 1942–May 9, 1945, of which 11,285,057 were [deaths] and 18,344,148 [wounded]." Colonel General G. F. Krivosheyev, cited by Lawrence G. Kelley in "The Soviet soldier in World War II: 'Death is but four steps away,'" p. 167, *Parameters, 27* (Winter 1997–98).

a nuclear monopoly (the U.S.S.R. detonated its first nuclear weapon on August 29, 1949, but did not have a true deployable nuclear capability until several years later), the U.S. greatly feared a Soviet lunge into central Europe—clearly a more critical strategic priority.[23] American strategists could not be sure whether or not the North Korean invasion was directed by Moscow to distract the U.S. and its allies. Given the intense ideological perspectives which dominated at the time, a judgment was made that communist states acted more or less monolithically and that an armed response was needed to contain further communist expansion. The Korean conflict ultimately absorbed much of the military capacity available against a peripheral, not central, strategic priority—a huge gamble. Its unsatisfying outcome, a negotiated armistice leading to a frozen conflict, reflected America's unwillingness to mobilize or commit totally to victory in a war not well understood or supported by the public. This "no win-no lose" approach would be seen again, with similar results.[24]

Many argued that the advent of nuclear weapons presaged the dislocation, or even negation, of grand strategy altogether. Through the 1950s, and despite the example of the Korean War, it was the declared policy of the United States to threaten a nuclear response to any attack. The international system settled into bipolarity, with two armed camps each capable of destroying the other absolutely, watched by "non-aligned" states who struggled to avoid co-option. Direct, armed confrontation between the U.S.S.R. and the U.S. seemed unthinkable for fear of uncontrolled escalation. Deterrence and containment became the means by which the ends of grand strategy were fulfilled. While powerful conventional forces were maintained, few strategists reckoned that the U.S. and its North Atlantic Treaty Organization (NATO) allies could prevail in a conventional war with the U.S.S.R. in the Central Region. Instead, nuclear systems at the tactical, theater, and inter-continental levels proliferated on

23 At this point NATO was only two months old and the German army had not been reconstituted, while the Red Army was far stronger than the U.S. Army.

24 At its high point, 327,000 U.S. troops served in Korea. U.S. war dead totaled 36,500.

both sides in an arms race only partially limited by arms control treaties. While the willingness of U.S. leaders to use nuclear weapons in Europe—to "trade Washington for Bonn"—was never certain, the consequences of miscalculation for either side were virtually unlimited, and deterrence in this sense proved remarkably stable. In only a single instance did the two superpowers approach the abyss—the Cuban Missile Crisis—and even then, the prospect of mutual destruction induced both to step back.

The long and painful experience of the Vietnam conflict reprised Korea and shared almost eerie similarities. Both featured ethnic populations, artificially partitioned. In both, the aggressor was a communist movement enabled and supported by China and the U.S.S.R. Both featured large, conventional forces fighting from protected sanctuaries. In both, the U.S. fought on the Asian mainland, far from the homeland and in a country with weak governance structures and a poorly developed infrastructure. And in both, U.S. air and seapower was unable to secure decisive battlefield results, even against a technologically-inferior opponent. Like Korea, Vietnam eventually consumed huge military resources at the expense of U.S. forces in Europe, miring the United States in a protracted, peripheral war with weak popular support. [25] In Vietnam, as in Korea, there were no direct threats to U.S. vital interests, only vague objectives to "resist communism" and to "maintain U.S. credibility."

Korea and Vietnam (and for that matter, smaller interventions such as Lebanon in 1958 and the Dominican Republic in 1965) took part against the backdrop of the Cold War and were clearly viewed in that light. For nearly five decades following the Second World War, national security concerns dominated the American political landscape as the United States engaged the Soviet Union in a worldwide struggle. For the first time in their history, Americans supported high defense expenditures in order to sustain large military forces in peacetime. Despite the painful experiences of the Korean and Vietnam conflicts, the United States never

25 U.S. troop strength in Vietnam eventually peaked at 536,000 in 1968. 58,000 U.S. service members were killed in the war.

faltered in its fundamental commitment to opposing Soviet expansion.[26] Internally or externally, there was little debate: deterrence or, failing that, fighting and winning our nation's wars went unquestioned as the defining tasks of the U.S. military.

Though far more dangerous, the Cold War was a simpler era in many respects. Our national security objectives were clear and unambiguous. Even at the height of the Vietnam conflict, the primary disagreement revolved around the nature of the struggle, not a questioning of the policy of containment. Sovereignty of individual states was paramount, tempered only somewhat by the moral force of international organizations such as the United Nations or, more concretely, by involvement in traditional security alliances such as NATO. The influence of non-state actors—whether nongovernmental organizations, private voluntary organizations, terrorist groups, drug cartels, international criminal syndicates, or others—was limited. In the main, national security imperatives were likely to prevail over other considerations in the strategic calculus.

All that changed when the Berlin Wall came down. Where superpower rivalry had previously inhibited the actions of ambitious regional powers and limited the influence of non-state actors, the collapse of the Soviet Union in 1991 led to immediate changes in the system that had governed international relations for over four decades. Overnight, the manifest threat ceased to exist. As a result, the United States and its allies were forced to adjust their strategic focus. At the same time, an increasingly interdependent global economy and emerging revolutions in information and communications eroded the concept of state sovereignty in fundamental ways. The result was a rise in international organized crime, quantum increases in international and domestic terrorism, ecological deterioration, disease, mass migration and refugee overflows, multiple outbreaks of ethnic and religious conflict, and a proliferation of failed states. These trends culminated in 9/11 and its painful and protracted aftermath.

26 The People's Republic of China, far weaker militarily and economically, played a role as a strategic balancer or counterweight but never approached superpower status during the Cold War.

The architects of the drawdown assumed, quite naturally, that a post-Cold War military would be far less busy in a world that would be more tranquil than before. Military forces were drawn down across the board. In one of the more interesting paradoxes of history, the end of the Cold War was followed, not by retrenchment or relaxation, but by a rapid increase in conflict and in our military commitments abroad. No longer driven by superpower rivalry, our national security policy evolved to advance our interests in a more fragmented, multipolar system largely defined by ethnic, religious, and cultural enmities as old as they were implacable. New challenges—economic, environmental, and factional as well as national, regional, and ideological—now confronted us in an international setting of greater complexity and variety.

These trends also fueled the rise of new actors on the international political landscape. The budget, influence, and level of activity of the United Nations and its many organizations increased substantially in the 1990s. Non-governmental organizations and private voluntary organizations became increasingly active, pursuing numerous, ambitious agendas in many different areas. Traditional national security concerns receded as the United States and other Western powers attempted to reap the dividends of peace. A fundamental shift took place, largely unnoticed, in the way many Americans viewed national security and the role of the armed forces in providing for the common defense.

The drawdown of the 1990s was wrenching. In a single decade, 700,000 U.S. military personnel spaces were eliminated (about one third of the active force), but the loss of combat forces was even more severe. In combat structure, the Army declined from eighteen active divisions to ten, the Navy went from 566 ships to 354, and the Air Force went from thirty-six to twenty fighter wings, an overall reduction of 45%. The defense budget in general terms dropped by 40%. In the midst of these changes, the military was asked to shoulder a heavier operational load. Stability operations in the Balkans, Haiti, and the Sinai in the 1990s stressed a force preoccupied with massive downsizing. Peace keeping, peace enforcement, and humanitarian assistance operations, as well as "theater engagement" missions, exploded. While the military had undertaken these types of missions throughout its history, the sheer number of deployments dwarfed those conducted in the past. Examples include refugee assistance in northern Iraq following the Gulf War; security and disaster relief efforts in Somalia; humanitarian aid to refugees in the Rwandan crisis area; restoration of democracy in Haiti; stability operations in Macedonia; peace enforcement operations to implement the Dayton Accords in Bosnia; and the Kosovo air campaign and later enforcement of the Military Technical Agreement in Kosovo.[27] More traditional combat or rescue missions in Panama, Southwest Asia, Liberia, Albania and elsewhere in the same time frame also stretched American forces and resources.

This dramatic turnaround in the international security environment could not help but impact our world in profound ways. Several trends heavily influenced American grand strategy since the Gulf War: the dramatic downsizing of U.S. military forces, their increasing use in non-traditional, non-combat missions, an increasingly polarized politics, and a prolonged period of economic distress and malaise. All are interrelated, and all have deeply affected the armed forces as instruments of national power, shaping U.S. strategy in important ways.

The 9/11 terrorist attacks struck to the heart of grand strategy as they represented the first large scale, direct attack on the homeland by an outside power since the War of 1812. Political unwillingness to confront the gathering threat and serious intelligence failures represented strategic failures for which the United States paid a high price. Following 9/11, defense spending increased substantially as the conflicts in Afghanistan and Iraq began and endured. Over a period of several years, the active Army grew from 470,000 to 548,000, with the Marine Corps expanding

27 Bosnia and Kosovo both evolved into lengthy and protracted commitments lasting many years.

from 158,000 to 202,000, while Air Force and Navy end strengths remained static or declined slightly. In keeping with Secretary of Defense Donald Rumsfeld's "Transformation" initiatives, significant investments were made in C4ISR systems and precision munitions, as well as in force protection enhancements such as up-armored wheeled vehicles. Nevertheless, legacy combat systems—planes, tanks, and ships, first delivered in the 1970s and early 1980s, remained the backbone of the military services (as they do today) while many next-generation programs were canceled or downsized.[28]

As with Korea and Vietnam, the post-9/11 era of conflict came to absorb much of our military effort and resources at the expense of other, more central security concerns.[29] In particular, ground forces were fully committed to the campaigns in Iraq and Afghanistan, leaving minimal active duty capacity for other contingencies such as the Korean peninsula.[30] Air and naval forces played much smaller roles. Over time, the Army in particular minimized its readiness for prolonged, state-on-state, high intensity conflict, shedding much of its armored, mechanized and field artillery force structure and focusing its combat training centers on counterinsurgency (COIN). The special operations community grew dramatically in size and capability in a single generation but could not play a decisive role in the counter-terrorism and counterinsurgency campaigns that defined the post-9/11 security landscape. With the U.S. effort in Iraq over and its Afghanistan venture winding down, it seems clear that neither will be seen retroactively as a clear-cut success, nor has

28 The F22 and F35 aircraft programs, intended to replace the F15 and F16, experienced significant cost overrun, production delay, and operational problems and were curtailed but protected in the Pentagon budget.

29 Despite a clear and compelling priority in Europe, both Korea and Vietnam eventually became the central focus for the U.S. Army and Marine Corps, leaving a much weaker land force in the Central Region. In much the same way, the Department of Defense arguably lost focus on nuclear deterrence and readiness, and on major theater war in scenarios like the Korean peninsula, accepting risk in order to focus on Iraq and later Afghanistan.

30 Large National Guard combat forces, up to twenty-eight brigades, exist in the force but require lengthy mobilization and would not be available to participate in a near-term crisis. Their sustained use also raises political questions which, in all but the most serious scenarios, are problematic.

the threat to the homeland from international terrorism been destroyed or eliminated.

At the conclusion of more than a decade of COIN, the United States finds itself repeating a familiar historical pattern. In the fiscal retrenchment that accompanies the end of every conflict (exacerbated by the economic collapse of 2008 and the Budget Control Act of 2011), active Army forces will bear the brunt of defense reductions, while the Air Force, Navy, and Marine Corps will be less affected.[31] Most U.S. ground and air forces have been repositioned to the continental U.S., while defense spending will decline over the next ten years by approximately 10% per year. At the same time, emerging, non-traditional threats such as cyber-attacks, weapons of mass destruction (whether chemical, biological or radiological) wielded by non-state actors, and international terrorism now crowd the security agenda. Increasingly, other threats such as narco-trafficking, illegal immigration, environmental degradation, demography, organized crime and even climate change are also cast as national security threats. What does this portend for American grand strategy?

The Ends of Grand Strategy

First, it is important not to confuse enduring, core strategic interests with others that are less central. The current security environment, described in the 2014 Quadrennial Defense Review (QDR) as "rapidly changing," "volatile," "unpredictable," and "in some cases more threatening" is certainly all those. Yet addressing this environment aligns comfortably with American grand strategy over time. Broadly speaking, U.S. vital or core interests remain remarkably consistent. They are: the defense of American territory and that of our allies, protecting American citizens at home and abroad, supporting and defending our constitutional values and forms of government, and promoting and securing the U.S. economy

31 Active Army forces, according to Pentagon sources, will fall to 420,000, the lowest level since before WWII. See *The Quadrennial Defense Review 2014*, p. ix (Washington D.C.: Government printing office), p. ix.

and standard of living. Virtually every strategic dynamic and dimension is encompassed in these four. Grand strategy is by no means confined to our military forces and institutions but is far broader, encompassing all forms of national power. That said, we must beware of attempts to define everything in terms of national security. Any discussion of grand strategy quickly loses coherence and utility when we do. [32] Grand strategy is fundamentally about security in its more traditional sense.[33]

Any assessment must begin with looking, first, at our security environment and, then, at threats to our core or vital interests, without either overestimating or undervaluing them. The international security environment is by now well understood and familiar. Raymond Aron's view of "a multiplicity of autonomous centers of decision and therefore a risk of war" holds true today.[34] The bipolar, traditionally Westphalian state system of the Cold War has given way to a more multi-polar system featuring a militarily- and economically-dominant (but not all-powerful) U.S.; a rising China and India; a resurgent Russia; an economically potent but militarily declining Europe; an unstable and violence-prone Middle East, wracked by the Sunni-Shia divide, economic and governance underperformance, and the Arab-Israeli problem; a proliferation of weak and failed states, particularly in Africa, the Middle East, and the Russian periphery; and empowered international and non-governmental organizations and non-state actors.[35] Terrorist organizations and

32 As one example of this tendency towards incoherence, Paul Doherty discusses the importance of "walkabout communities" as part of a "new grand strategic construct" in "A New U.S. Grand Strategy," *Foreign Policy*, January 9, 2013.

33 This trend is driven in part by a desire to access defense budgets to fund programs not traditionally considered as defense-related. Stanley Hoffman put it succinctly as far back as 1987: "There has been a trend towards indefinite extension of U.S. interests. "National security" is considered to be everywhere and constantly at stake." See Janus and Minerva, *Essays in the theory and practice of international politics*" (Boulder CO: Westview Press, 1987), p. 316.

34 Raymond Aron, *Peace and War* (Paris: Calmann-Levy, 1962), p. 28.

35 As recently as 2002, William C. Wohlforth argued that "the balancing imperative . . . will not soon dominate great powers' strategic choices in today's novel unipolar system." In fact, though the U.S. remains unquestionably the preponderant world power, great powers like Russia, Iran, and China often combine to limit or deflect U.S. strategic choices in a classic balance of power formulation. Wohlforth, "US strategy in a unipolar world," in *America unrivaled: The future of the balance of power*, G. John Ikenberry (Ed.), (Ithaca NY: Cornell University Press, 2002), p. 117.

international organized crime are far more significant than formerly, enabled by global communications and information flows. In absolute terms, the world is safer, as the prospect of nuclear, mutually assured destruction and world war costing millions of lives seems relegated to the past. Yet most societies feel threatened and insecure, while conflict, if more low-level, remains endemic.

In this regard, we often see references to "asymmetric" threats, posed to "thwart U.S. conventional military advantages."[36] While factually true—weaker states find it largely impossible to match U.S. power symmetrically—this characterization can be misleading. It is just as accurate to cast asymmetric threats as less capable offsets employed by weaker powers who cannot match American preponderance. A persistent tendency to inflate the dangers of insurgency, terrorism, "niche" technologies and so on can distort threat assessments in unhelpful ways. Asymmetric threats deserve careful consideration but should not be exaggerated.

The broad threats which face us have deep roots but have also evolved over time. In order of importance, they can be summarized as:

- Use of weapons of mass destruction (WMD) against the homeland. These could be nuclear, chemical, biological, cyber, or explosive/kinetic in nature (such as the 9/11 attacks) delivered by either state or non-state actors. Single or multiple attacks causing huge mass casualties could lead to partial or complete economic collapse and loss of confidence in our governance structures, imperiling our standard of living and way of life in addition to loss of life.[37]

36 Julianne Smith and Jacob Stokes, "Strategy and statecraft: An agenda for the United States in an era of compounding complexity," *Center for a new American security*, June 2014, p. 9.

37 The official Department of Justice definition of WMD includes nuclear, chemical, biological and radiological weapons only. The term is used more broadly here to include events like the Oklahoma City bombing and the 9/11 attacks on the World Trade Center and Pentagon, as well as potential cyber events that could cause large scale loss of life. See Seth Carus, "Defining Weapons

- Economic disruption from without. The financial crisis of 2008 was largely self-induced, but the health and stability of the U.S. economy can also be affected by the actions of foreign powers. Saddam Hussein's invasion of Kuwait, which jeopardized the international economic order by threatening the free flow of oil from the Persian Gulf, is an example. A major cyber-attack against the financial sector or closure of the Strait of Tiran or the Straits of Malacca by a hostile power could be another.[38] Any major disruption to the global economy, which depends upon investor confidence as much as the free flow of goods and energy, can have catastrophic consequences for the U.S., and American presidents have repeatedly shown a willingness to use force to ensure access to markets, free trade, and economic stability.

- The rise of a hostile peer competitor. For centuries, Great Britain aligned against the rise of any power able to dominate the European landmass and upset the balance of power. The United States did the same in opposing Germany in WWI, Germany and Japan in WWII, and the Soviet Union in the Cold War. The U.S. "Rebalance to Asia" and opposition to Chinese territorial moves in the East and South China Seas can be seen as an attempt to counter the rise of China in a manner consistent with long-standing U.S. grand strategy. A peaceful, non-hostile peer nation or grouping of nations (such as the European Union) poses no strategic threat to the United States. An authoritarian great power, possessed of both military and economic means and an apparent desire to enlarge and expand them, could in time pose a direct, existential threat to American national security. American

of Mass Destruction," *Occasional Paper 8*, Center for the Study of Weapons of Mass Destruction (Washington, D.C.: National Defense University Press, January 2012).

38 One quarter of the world's traded goods and 25% of the world's oil supply carried by sea passes through the 2.8 km wide Phillips Channel south of Singapore.

grand strategy has traditionally opposed such powers and would in all likelihood do so again.[39]

- Direct challenges to key allies. Alliances like NATO and bilateral security arrangements with close allies like Japan and South Korea constitute solemn commitments that extend American power and influence globally. Cooperation with allies adds their military forces to ours and secures forward basing and other rights we need to secure U.S. interests around the world. We do not enter into them altruistically, but because they serve U.S. interests. To preserve international stability and deter conflict, they must be honored. Failure to do so in one case, such as an attack on Japan or South Korea, would call into question our commitment to all such commitments. This could compromise, perhaps fatally, our system of alliances and treaties worldwide.[40] U.S. leaders can be expected to act decisively when close allies are directly threatened.

There are, of course, other threats of concern to national security practitioners which fall below this threshold. An attack on a U.S. embassy, the kidnapping of U.S. citizens abroad, or the pirating of U.S.-flagged vessels on the high seas are examples. U.S. political leaders might also contemplate the use of military force under the evolving doctrine of "Responsibility to Protect" (R2P), as in the case of Somalia in 1991 and Libya in 2011, or when national pride has been touched (as in the Mayaguez or Pueblo incidents). However, these by definition do not engage grand strategic objectives, and statesmen assume risk when

39 This does not necessarily mean through military confrontation. In the case of China, for example, the fact that many neighbors with substantial military establishments (Japan, Russia, India, South Korea, Vietnam, Taiwan, Indonesia, Malaysia, Singapore) can check possible military expansion when acting in concert suggests that U.S. security and economic assistance and diplomacy may be the primary venues to constrain the rise of a potentially-expansionist China.

40 Treaties and alliances are means to an end, not an end in themselves, but preserving them is clearly a core interest for which the U.S. will use force if necessary.

treating them as though they do, primarily because strong and sustained public support is less assured.[41]

Similarly, promoting democracy and human rights abroad are often touted as national security or foreign policy "imperatives."[42] While consistent with American political culture and ideology, in practice, these are highly case specific. When consonant with the framework and principles of U.S. grand strategy, the U.S. may act, but, more often, a pragmatic realism governs.[43] The long nightmare in Syria—with its tragic loss of life, accelerating regional instability, mounting extremism and terrorist involvement, and massive human rights violations on all sides—would seem to be a classic case calling for military intervention. Yet there is no UN or NATO mandate, no strong reservoir of public support for military action, no appetite for intervention among our allies and partners, and no desire to dispute the agendas of Russia, China, and Iran in Syria, at least for the time being. With no direct threat to the homeland, U.S. citizens or allies, or the US economy, the prospects for large-scale military intervention at present seem low, despite the humanitarian tragedy unfolding.

The picture changes as we look at the expansion of the Islamic State in Iraq and Syria (ISIS) into northern Iraq and the establishment of the "Islamic caliphate." This development threatened to destroy the Iraqi state, destabilize the entire region, and provide a platform and safe haven to launch terrorist attacks across the region and against the West and the homeland. The indiscriminate killing of hundreds of civilians,

41 "Ironically, a war fought in the name of high moral principle intensifies violence and is more destructive of political stability than a war based on national interest." Kenneth Thompson, *Masters of international thought* (Baton Rouge: Louisiana State University Press, 1980), p. 153. An almost tragic example is provided by Secretary of State Dean Rusk at the height of Vietnam, who was quoted as saying "the United States cannot be secure until the total international environment is ideologically safe." Cited in Layne, "Kant or cant: The myth of the democratic peace," *International Security*, *19*(2), Fall 1994, p. 46.

42 The 2014 QDR lists "respect for universal values at home and around the world" as one of four "core national interests." *QDR 2014*, p. 11.

43 See R. D. Hooker, Jr., "US policy choices during the Rwandan genocide," unpublished paper, National War College, 2003.

particularly minorities, and the videotaped beheading of a captured American journalist in August of 2014 moved this threat to center stage. Working largely through Kurdish proxies enabled by American airpower and special operations forces (SOF), the U.S.-led coalition succeeded in destroying the geographic ISIS caliphate and driving its survivors underground over some three years.

The crisis in Ukraine presents a different case study. While the prospect of committing U.S. forces to defend Ukraine following the seizure of the Crimea is low, the post-war security architecture in the Euro-Russian space, so carefully constructed for a generation, has been thrown over. The North Atlantic Council voted to defer NATO membership to Georgia and Ukraine and did not station NATO troops in the new member states, largely out of deference to Russian security concerns. These confidence-building measures notwithstanding, Russia sent troops into Georgia in 2008, where they remain today.[44] In particular, the 1994 Budapest Memorandum, under which Ukraine agreed to surrender its nuclear weapons in exchange for Russian guarantees of its territorial integrity, has been seriously compromised, along with the Organization for Security and Cooperation in Europe (OSCE) and, apparently, the NATO-Russia Council. Russian forces, having seized Crimea, have intervened with tanks and artillery inside Ukrainian territory. Concerns by NATO members, especially the newer ones located in Eastern Europe and the Baltic States, are mounting as Russian leaders assert the right to "protect" ethnic Russian minorities in neighboring countries.

This scenario presents a different challenge to American grand strategy. Should Russia seize more Ukrainian territory, NATO's Baltic members could very possibly come under threat, an altogether different matter.[45] Russian subversion or military action in Latvia, Lithuania, and

44 In addition to troop deployments to Abkhazia and South Ossetia, "breakaway" regions which remain part of sovereign Georgian territory, Russia maintains 5,000 troops inside Armenia to ensure that neighboring Azerbaijan does not reclaim Nagorno-Karabakh, sovereign Azerbaijan territory occupied by ethnic Armenians.

45 Latvia, Lithuania, and Estonia are 100% dependent on Russian natural gas and have large

Estonia may be deterred by NATO's Article 5 guarantee, and U.S. leaders and their allied and partner counterparts will work hard to dissuade any thought of further aggression through energetic diplomacy and severe economic sanctions. Still, should Russia reprise its Crimea land-grab in the Baltic states, it is more likely than not that the U.S. will respond militarily under the Washington Treaty and encourage its NATO allies to do the same. Direct confrontation with Russia, still a major nuclear and conventional power, may seem unthinkable. Yet failure to honor our treaty obligations to NATO would mean the virtual collapse not only of the alliance but also of our security relationships around the world. Such a loss of global reach and influence would negate U.S. grand strategy altogether. For that reason, however much against its will, the U.S. will in all likelihood confront Russia should a NATO member be attacked or directly threatened.

These and similar examples raise the question of whether or not the United States consciously pursues an imperial or hegemonic grand strategy. Many scholars, both domestic and foreign, explicitly or implicitly assert that it does.[46] On the one hand, the U.S., along with other great powers, seeks to provide for its own security by maximizing its power relative to potential and actual adversaries, within limits imposed by its domestic politics. Its political and military leaders are constrained in attempting to balance what Aron called an ethics of responsibility—the pragmatic reality of an international politics which cannot and does not ignore the role of force—and an ethics of conviction, which is normative and classically liberal in seeking accommodation and an absence of conflict where possible.[47] It is thus true that American power, and particularly military power, is often employed to secure and advance American interests. On

ethnic Russian populations. Without NATO, their continued independence is probably unlikely. See Theresa Sabonis-Helf, "Energy security: Strategic questions and emerging trends," presentation to NATO National Representatives, National Defense University, April 11, 2014.

46 Andrew Bacevich is a leading critic of American "imperialism." See *American empire: The realities and consequences of U.S. diplomacy* (Cambridge, MA: Harvard University Press, 2002).

47 Thompson, p. 175.

the other hand, U.S. interventions are marked by an absence of territorial aggrandizement or forced extraction of natural resources. Typically, huge sums are spent on development and infrastructural improvements. On its own or when asked (as in the Balkans; in Somalia, Haiti, and Panama; and in Iraq), America usually withdraws and goes home. Even close allies remain free to opt out of military ventures, as seen in the invasion of Iraq in 2003 and in Libya in 2011.

The net effect has been to bring into being, largely if not entirely through America's own efforts, a rules-based international and economic order that has widely benefited much of the world:

> It falls to the dominant state to create the conditions under which economic interdependence can take hold (by providing security, rules of the game, and a reserve currency, and by acting as the global economy's banker and lender of last resort). Without a dominant power to perform these tasks, economic interdependence does not happen. Indeed, free trade and interdependence have occurred in the modern international system only during the hegemonies of Victorian Britain and postwar America.[48]

These are the actions of a preponderant, but hardly of a classically imperialist, power. If the U.S. is imperialist, it appears to be so in an historically benign way; if hegemonic, in a heavily qualified one.[49]

The Means of Grand Strategy

The "means" of grand strategy are similarly enduring over time. Fostering strong alliances and bi-lateral security arrangements,[50]

48 Layne, "Rethinking American grand strategy," p. 15.

49 "[t]he Western postwar order has also been rendered acceptable to Europe and Japan because American hegemony is built around decidedly liberal features." G. John Ikenberry, "Institutions, strategic restraint and the persistence of the American postwar order," *International Security, 23*(3), p. 49.

50 "The United States leads a global alliance system of more than sixty partner states that

maintaining a strong and survivable nuclear deterrent; fielding balanced, powerful, and capable military forces; dominant in each warfighting domain, that can project and sustain military power globally and prevail in armed conflict; and providing intelligence services that can ensure global situational awareness and provide strategic early warning are basic components. They are intrinsically linked to a powerful economy and industrial base, advanced technology, an extensive military reserve component, an educated and technically skilled population fit for military service,[51] and a political system based on classically liberal democratic values that is able to make clear and sustainable policy and resource decisions.[52]

In important ways, these tools and capabilities are, or are perceived to be, eroding. The U.S. economy, still the largest in the world, has only now fully recovered from the 2008 crisis. America's traditional reliance on forward presence and forward deployed forces, another strategic linchpin, has also declined since the end of the Cold War. Few combat forces remain in Europe, only a single ground combat brigade is based in Korea, and ground combat troops based in the Middle East are few and advisory in nature. Naval forward presence has also been scaled back in

collectively account for almost 80% of global GDP and more than 80% of global military spending between them." Michael O'Hanlon, *Budgeting for hard power: Defense and security spending under Barack Obama* (Washington, D.C.: Brookings, 2009), p. 24.

51 The General Accounting Office reports that 16.2 million males aged 18–25 are registered for selective service. However, only one in four are eligible for military service, severely limiting the pool of prospective recruits. The rest are disqualified for obesity, other physical issues, lack of a high school diploma, or criminal records. See the prepared statement of Dr. Curtis Gilroy, director for accessions policy, Office of the Undersecretary of Defense for Personnel and Readiness, before the House Armed Services Subcommittee "Recruiting, retention, and end strength overview," March 9, 2009.

52 The U.S. industrial base remains a world leader, second only to China as of 2014 according to the United Nations Statistics Division. Ship building remains a strong industry: "Currently there are 117 shipyards in the United States, spread across twenty-six states, that are classified as active shipbuilders". "The economic importance of the U.S. shipbuilding and repairing industry," *Maritime Administration*, May 30, 2013, p. 3. U.S. steel production has declined as a percentage of global market share since 1947, when the U.S. produced 60% of the world's steel, but remains a world leader. The United States produced 87 million tons of steel in 2013, ranking fourth in the world. (This contrasts with 40 million tons dedicated to military production in 1943, the year of greatest manufacturing output in WWII) See *World Steel Statistics Data 2013*, World Steel Association, January 23, 2014, and Gropman, p. 137. For a contrary view, see O'Hanlon, "The national security industrial base: A crucial asset of the United States whose future may be in jeopardy," Washington, D.C.: Brookings, February 2011.

the post-Cold War era, as the size of the fleet has declined.[53] On the alliance front, relations with NATO allies have been damaged by the Rebalance to Asia, widely perceived as a devaluation of Europe by U.S. leaders, and by Secretary Gates's stern speech in June of 2011, which castigated European allies for failing to meet targets for defense spending.[54] President Obama's "leading from behind" stance in Libya, the pullout from Iraq, the pending withdrawal from Afghanistan, and inaction in Syria are interpreted by some as evidence of a disinclination to engage globally in the interests of international stability, though others see prudent and measured restraint.

The use of "soft power" also deserves consideration in this discussion.[55] Described by its progenitor as "the ability to influence the behavior of others to get the outcomes you want,"[56] soft power is concerned with development aid, cultural influence, the power of example, and others forms of suasion which are not coercive or easily directed. Theorists disagree on whether or not soft power should be considered as part of the strategist's arsenal. Diplomacy, for instance, may lack utility when divorced from the military and economic power of the state; the artfulness of the discussion may be useful but will not be decisive absent hard power. On balance, though the ability of soft power to influence adversary behavior for good or ill is probably incontrovertible, it is not easily deployable or even controllable.[57] To that extent, it is an important factor that nevertheless falls outside the realm of grand strategy as traditionally understood and practiced.

53 Currently, 36% of the Navy's operational assets are classed as "globally deployed", including two of the Navy's eleven fleet carriers, with a third based in Japan. At least two Amphibious Ready Groups with embarked Marines are also always at sea. See Admiral Jonathan Greenert, Testimony Before the House Armed Services Committee on the FY2015 Navy Posture, March 12, 2014.

54 Thom Shanker and Steve Erlanger, "Blunt U.S. Warning Reveals Deep Strains in NATO," *The New York Times*, June 10, 2011.

55 "Soft power is not about influence or persuasion—it attracts." Harry R. Yarger, *Strategy and the National Security Professional* (London: Praeger, 2008), p. 74.

56 See Joseph Nye, *Soft power: The means to succeed in world politics* (New York: Public Affairs, 2004).

57 Chris Schnaubelt, "The illusions and delusions of smart power," in *Towards a Comprehensive Approach: Integrating Civilian and Military Concepts of Strategy* (Rome, Italy: NATO Defense College Forum Paper, March 2011), p. 24.

While U.S. determination to act forcefully in support of the international order may be more open to question, and while U.S. economic and military power may not be as dominant as formerly, in absolute terms, the United States remains by far the preponderant power in the world. Possessed of great actual and potential strengths, in hard power, the United States is unequalled. Nevertheless, coherent and effective political direction is the essential pre-condition to strategic success. Since the end of the Vietnam War, mounting conflict between the legislative and executive branches—spurred by a fractious polarization of American politics—has reached alarming proportions. Repeated wars have led to a concentration of the war power in the executive branch, arguably resulting in more frequent uses of force that may not command public support. Unquestionably, a healthy and stable set of political arrangements that provides for effective sharing of power, while ensuring popular backing, is essential.[58] When lacking, successful strategic execution is at risk.

The Ways of Grand Strategy

How America addresses direct threats to its core or vital interests over time is the essence of grand strategy. Typically, its solutions are not new, although the technologies employed often are. The first principle is to meet the threat as far from the homeland as possible. Thus, since the end of the Second World War the U.S. has established bases, positioned forces and stockpiled weapons and munitions around the globe, buttressed by economic and development assistance, exercises, formal treaties, coalitions of the willing and alliances.[59] (Counter-proliferation may also be seen in this light.) While U.S. ground forces have largely come home, and key installations like Torrejon Air Base in Spain, or Clark Air Base and Subic

58 Effective civil-military relations is also, of course, a *sine qua non* of successful strategy. Despite much hyperbolic academic criticism, the U.S. is well equipped in this sphere. See Hooker, "Soldiers of the State: Reconsidering American Civil-Military Relations," *Parameters*, Winter 2003–2004.

59 The Department of Defense maintains prepositioned stocks both ashore and afloat in strategic locations worldwide to support the deployment of forces for contingency operations. Key sites are Japan, Korea, Italy, Qatar, Kuwait, and Diego Garcia.

Bay in the Philippines were closed after the Cold War, America's network of overseas bases, airfields, and alliances, as well as forward deployed air and naval forces, is still extensive. America's ability to project power globally and sustain its forces virtually indefinitely remains unmatched. U.S. satellites survey the globe and monitor adversary communications continuously. Though smaller than during the Cold War, the U.S. strategic nuclear arsenal is survivable, redundant, and accurate, providing an absolute nuclear deterrent against any adversary.[60]

Next, the U.S. prefers to meet serious threats using different tools at once, in theory reserving military force for last and relying on intelligence, diplomacy, forward presence, and its economic power to forestall, deflect, or defuse security challenges.[61] Still, U.S. military power is awesome. Its strength across the warfighting domains, supported by an unmatched ability to project and sustain military forces far from the homeland, remains far ahead of the rest of the world.[62] Whenever possible, the U.S. prefers to address threats in tandem with allies, partners, or like-minded states, working through international organizations like the UN or NATO and conducting pre-conflict engagement and "shaping" operations on a large scale. Yet when vital interests are engaged, the U.S. will act unilaterally if necessary.[63] Preemption to disrupt or prevent imminent

60 The most recent arms control agreement with Russia, signed by President Obama and Russian President Medvedev on April 8, 2010, agreed to reduce the number of active nuclear weapons from 2,200 to 1,550.

61 *Quadrennial Defense Review 2014* (Washington: GPO, March 4, 2014), p. 11.

62 For example, in seapower alone the U.S. lead is staggering. The U.S. Navy operates eleven large aircraft carriers, all nuclear powered; no other country has even one. The U.S. has fifty-seven nuclear-powered attack and cruise missile submarines—again, more than the rest of the world combined. Seventy-nine Aegis-equipped surface combatants carry roughly 8,000 vertical- launch missile cells, outmatching the next twenty largest navies. All told, the displacement of the U.S. battle fleet exceeds the next thirteen navies combined, of which eleven are allies or partners. Cited in Secretary of Defense Robert Gates' prepared remarks to the Navy League, National Harbor, Maryland, May 3, 2010. The United States Marine Corps alone is larger and more capable than the ground and air forces of all but a few nations. See *The military balance* (London: The International Institute for Strategic Studies), March 2014.

63 "The United States will use military force, unilaterally if necessary, when our core interests demand it—when our people are threatened; when our livelihoods are at stake; when the security of our allies is in danger." President Barak Obama, Commencement Address at the United States Military Academy, May 28, 2014.

threats falls well within America's grand strategic calculus.[64] Prevention—the use of force to defeat threats before they become imminent—has, on the other hand, far less provenance.

As the preponderant global power, the U.S. attempts to shape the international security environment to prevent or ward off security challenges where it can.[65] When it cannot, and when significant or vital interests are engaged, military force often comes into play. Since the end of the Second World War, the United States has used military force many times, with varying success, to protect, secure, or advance its security interests.[66] When military force was used, the record of success or failure is illustrative when viewed in light of the grand strategic framework described above. In the previous century, the U.S. has experienced clear success when the threats to vital interests were unambiguous; when the response enjoyed strong support from the public and Congress; when overwhelming force was applied; when strong allies participated; and when the strategic objective was well understood.[67] Both World Wars, the Cold War, and the Gulf War are examples. In cases where the direct threat to U.S. vital interests was less clear, overwhelming force was not applied, public and congressional support was not strong or sustained, and the strategic objective was unclear, defeat or stalemate ensued. Korea, Vietnam, Beirut, Somalia, Iraq, and Afghanistan are, of course, the relevant examples here. In some cases (the Dominican Republic, Grenada, Panama, Haiti, Kosovo), the desiderata listed above did not fully

64 *The National Security Strategy of the United States*, published on September 17, 2002, p. 15.

65 For a more detailed discussion of American preponderance and its strategic implications, see Layne, p. 9.

66 The list of large-scale combat or "peace enforcement" actions alone is extensive and includes Korea (1950), Lebanon (1958), the Dominican Republic (1965), Vietnam (1955–1975), Beirut (1981), Grenada (1981), Panama (1989), the Gulf War (1991), Somalia (1992), Bosnia (1996), Kosovo (1999), Afghanistan (2001), and Iraq (2003). On average, the U.S. has deployed a division or larger force every six years since 1950.

67 Both Secretary of Defense Caspar Weinberger and Chairman of the Joint Chiefs General Colin Powell promoted similar views on when and how to use force, espousing a conservative, "last resort" philosophy stressing overwhelming force and clear objectives and emphasizing decisive results. Weinberger explained his in a speech entitled "The Uses of Military Power" delivered before the National Press Club in Washington, D.C., on November 28, 1984.

apply, but weak opposition and overmatching force led to early success, forestalling loss of public support or stagnation of the conflict.[68]

These historical lessons are compelling and deserve careful and objective study. American political leaders have not always recognized these principles and have certainly not always applied them. Their apparent jettisoning by both Republican and Democratic administrations following the Gulf War has come with a heavy price. America's successes in war, and in deterring war, have resulted at least as much from an industrial and technological superiority, employed en masse by competent political and military institutions, as from any other factor.[69] This superiority is best translated into battlefield and campaign success by synergistically applying air, space, sea, cyber, and land power in time and space to achieve decisive objectives that see through and beyond the end of combat operations. One-service or one-dimensional applications of force have repeatedly failed of their promise to deliver strategic victory.

Relatedly, political leaders and strategists should be mindful of *strategic culture*, that mélange of history, tradition, custom, world view, economy, sociology, political systems, and more that largely shapes how nations fight and for what causes. There may be no agreed-upon American theory of war, but when based on concepts of joint and combined warfare, an "American Way of War" surely obtains mass, firepower, technology, strong popular support, and a focus on decisive and clear cut outcomes.[70] "Good wars" have historically followed this pattern. "Bad wars" have not. While the analogy can be taken too far, it captures central truths that

68 Author and intellectual Max Boot attempted to argue in 2002 that "small wars" fought for less precise objectives could advance important, if not vital, interests and represented something of a future trend. On the whole, such thinking has been discredited by Iraq and Afghanistan. See Boot, *Savage wars of peace* (NY: Basic Books, 2002).

69 "By and large, the virtues of American civilization have not been the military virtues and this has been reflected in American military performance." Samuel P. Huntington, "Playing to win," *The National Interest*, Spring 1986, p. 10.

70 Weigley is the principal exponent of this view. For a contrasting view, see Antulio Echevarria's "Toward an American way of war," *US Army War College Strategic Studies Institute*, March 2004.

should inform our strategic calculations.[71] Strategic culture is real and powerful, whether acknowledged or not.[72]

The Way Ahead

As we assess a complex security environment, our historical experience provides useful and helpful context and guideposts to understanding the present, even when security threats are harder to define and address, as in the case of cyber-attacks.[73] U.S. forces are also held to standards increasingly difficult to guarantee; the prospect of even minimal casualties to our own forces or to civilians (however unintentional) or unintended environmental damage now colors every decision in the age of the 24-hour news cycle. On balance, traditional military security concerns often seem less paramount. Absent a clear and present danger, humanitarian considerations, environmental issues, and resource impacts and scarcities compete strongly with military factors in policy deliberations. In the meantime, non-state actors are increasing their power and influence to affect policy changes across a wide spectrum of issues, many of which directly affect the ability of U.S. military forces to carry out their missions.[74]

In the last generation, we have often seen the face of the future reflected in the bitter divisions of the past, in failed states, in emerging democracies and in nations stuck in transition between authoritarian and democratic systems. A persistently uncertain and unstable international security environment places a premium on U.S. leadership. As the only

71 "We make war the way we make wealth." Alvin and Heidi's Toffler, *War and anti-war: Survival at the dawn of the 21st century* (New York: Little, Brown, and Company, 1993), p. 2.

72 See Hooker, "The strange voyage: A short précis on strategy," *Parameters*, Winter/Spring 2013, p. 62.

73 "I believe the most pressing threat facing our country is the threat from cyber attacks. The daily occurrences of attacks are damaging on a variety of levels and they are not only persistent and dangerous, the likelihood of serious damage to our national security is very real." Lieutenant General Mike Flynn, Director, Defense Intelligence Agency, *Statement Before the Senate Armed Services Committee, United States Senate*, April 18, 2013.

74 The international treaty banning landmines in 1999 and the International Criminal Court, established in 2002, are apposite examples. The U.S. is not a party to either.

remaining global power and as a coalition leader in organizations like NATO, the U.S. is uniquely positioned to influence world affairs in ways that benefit not only the United States but also the international community as a whole.[75] The prudent use of American military power, in concert with the economic, political, and diplomatic instruments of national power, remains central to attempts to shape the international environment and encourage peace and stability wherever important U.S. interests are at stake.[76] As George Kennan put it:

> We have learned not to recoil from the struggle for power as something shocking or abnormal. It is the medium in which we work . . . and we will not improve our performance by trying to dress it up as something else.[77]

Much of the prevailing academic discussion, on the other hand, distracts or frustrates practitioners. One leading theorist offered presidents a choice from among strategies of "neo-isolationism, selective engagement, cooperative security, primacy, or enlargement and engagement."[78] Another proposes "strategic restraint, offshore balancing, forward partnering, selective engagement or assertive interventionism" as strategic alternatives.[79] Others argue for regional priorities (Asia/Pacific, or the Middle East, or Europe), or threat-based priorities (weapons of

75 As the only state able to project and sustain military forces globally, the U.S. retains this status today; the rise of China will not see an equivalent capability for years to come. Andrew J. Nathan and Andrew Scobell, *China's search for security* (New York: Columbia University Press, 2012), pp. 312–315.

76 Theorists sometimes cite the maxim that "everybody's strategy depends on everyone else's." This must be the case for weaker or comparable powers. In its current position of preponderance, though its power has definite limits, the United States seeks whenever possible to impose its strategy on adversaries, and not to be imposed upon. All states would behave so if they could. See Kenneth Waltz, *Man, the state and war* (New York: Columbia University Press, 1954), p. 201.

77 George Kennan, *Memoirs* (Boston: Little and Co., 1967), p. 494.

78 Barry R. Posen and Andrew L. Ross, "Competing visions for US grand strategy," *International Security*, *21*(3), Winter 1996/1997, pp. 5–53.

79 F. G. Hoffman, "Forward partnership: A sustainable American strategy," *Orbis*, Winter 2013, p. 24.

mass destruction (WMD), cyber, insurgency), or for capabilities-based strategies (for example, the Maritime Strategy of the 1980s). Each approach offers useful perspectives, but true grand strategy looks beyond these choices, orienting on American strengths and American interests to address the global challenges of the moment in a larger framework of diplomacy, economic strength, military power and global leadership. Presidents do not really have the choice to embrace isolationism, ignore alliances, eschew engagement or ignore important regions of the world. The administration may highlight the Rebalance to Asia as its top priority, but potential conflict in the Arabian Gulf, another WMD attack on the homeland or Russian military action against the Baltic States, as examples, will immediately become the pressing, consuming challenge and will remain a critical priority until resolved.

It is also useful to note that the formerly sharp distinction between the military instrument and others has become blurred. The definition of "national security" is now more expansive, encompassing a great domain of "homeland defense," with dozens of civilian agencies and large military organizations (such as U.S. Northern Command) intimately linked with and often working in subordination to other civilian entities. Even in conflict zones, tactical formations engaged in daily combat can find themselves with scores of embedded civilians representing civilian departments.[80] Informational technologies and a more globalized threat, able to strike from remote and under-developed locations with great effect, now force a greater degree of synergy and interoperability between military and non-military organizations than ever before. These trends will continue on a trajectory towards ever-greater civil-military integration, particularly in the intelligence, cyber, acquisition, logistics and consequence management realms.

80 For example, Regional Command-East in Afghanistan in 2010 included a Senior Civilian Representative from the Agency for International Development, of equal rank to the division commander and empowered to co-sign his operational orders. She was supported by more than one hundred civilian staff.

Taking the long view, and acknowledging the strong impact of new technologies and threats, the framework of American grand strategy as described here will remain relevant and current for decades to come. The international security environment will remain anarchic and uncertain, with the state mattering more than supra-national organizations even as non-state actors of many kinds proliferate. Conflict will remain endemic, and state-on-state conflict will recur. WMD attacks against the homeland will be attempted, and may be successful. Pressures to intervene—in the Middle East, in Africa, in Eastern Europe and perhaps even in East Asia—will persist or surface anew. Strategic "shocks"—unanticipated crises requiring strategic responses—will be more the norm than not.[81] None of this is new, unique or even more dangerous than in the past.

Strategists must accordingly consider and refine the ways and means by which our traditional and enduring interests may best be defended. Along the way, a certain humility is helpful; as Kissinger wrote, "the gods are offended by hubris. They resent the presumption that events can be totally predicted and managed."[82] At its best, grand strategy is not always or fundamentally about fighting or the military application of force, but rather an appreciation of its potential, along with the other instruments of power, in the mind of the adversary. President Reagan's role in bringing about an end to the Cold War is the classic example. In this sense, effective grand strategy may often preclude the need to resort to force. To achieve this, the involvement of society in its own national defense, a strong, stable and globally networked economy, an effective domestic politics that can make rational decisions over time in support of national security, and the promotion of values that invite support and consensus at home and abroad will count for much. So, too, will balanced and capable military forces, sized and able to operate globally and in concert with civilian counterparts, international organizations, allies and partners. The decision when and if to use force should never be approached casually,

81 Strachan, p. 1285.

82 Cited in Brands, p. 100.

emotionally, or halfheartedly, but rather soberly, analytically and with a whole of government and whole of society intention to prevail. There should never be doubt that when core interests are engaged, the U.S. will bring the full weight of its power to bear and will persist until success is achieved. On these foundations will rest an effective U.S. grand strategy far into the future.

2 The Strange Voyage

A Short Précis on Strategy

Practitioners know well that actual strategy falls well short of the clean, precise prescriptions of theory. Often, strategy is a leap into the unknown, buffeted by chance, miscalculation, and unanticipated consequences. Here, the author provides context and insights into strategy as it exists in the real world.[1]

Try as we might, war and armed conflict remain at the center of international relations and state policy. Success in war requires many things, but surely effective strategy must top the list. Why is making good strategy so hard? Because it is, perhaps, the most difficult task facing senior leaders in any government. Despite a wealth of sources and millennia of useful historical examples, sound strategic thinking more often than not eludes western democracies. Why?

History has a way of making strategy look simple and even inevitable. In the common narrative, for example, Pearl Harbor forced America into WWII, the U.S. adopted a "Europe first" approach, went to full mobilization, led victorious coalitions to smash the opposition, and then won the peace. The reality was very different, the outcome at the time far from certain, and the costs required far higher than expected. Strategic reality is more accurately captured by Winston Churchill's term "the

1 Revised from an essay that originally appeared in *Parameters*, Winter 2012/2013.

strange voyage."[2] So often begun with confidence and optimism, strategic ventures often end in frustration and indecisive outcomes.

Good strategy begins with basic questions. What are we trying to do? How much will it cost? How should we use what we've got to achieve the aim? The questions are simple. But answering them—thoughtfully, comprehensively, honestly, and dispassionately—is by far the exception to the rule. Failing to frame the problem correctly at the outset may be the most common, and disastrous, strategic error of all.

The first minefield is one of definition. Students, theorists, and practitioners of strategy face a bewildering range of competing and confusing terms. Thus we find National Security Strategy, National Defense Strategy, National Military Strategy, grand strategy, coalition strategy, regional strategy, theater strategy, and campaign strategy—to name a few. Where does one end and the other begin? Do they overlap? Or are some just synonyms? The word "strategy" comes from the Greek *strategos*, the word for "General." Classically, strategy was quite literally "the Art of the General." Webster defines strategy as "the science and art of employing the political, economic, psychological, and military forces of a nation or group of nations to afford the maximum support to adopted policies in peace or war." The military prefers "a prudent idea or set of ideas for employing the instruments of national power in a synchronized and integrated fashion to achieve theatre, national, and/or multinational objectives."[3] Clausewitz defined strategy as "the art of the employment of battles as a means to gain the object of war." The great Moltke used "the practical adaptation of the means placed at a general's disposal to the attainment of the object in view," famously observing that strategy is most often "a system of expedients." Liddell-Hart favored "the art of

2 "Never, never, never believe any war will be smooth and easy, or that anyone who embarks on the strange voyage can measure the tides and hurricanes he will encounter. The statesman who yields to war fever must realize that once the signal is given, he is no longer the master of policy but the slave of unforeseeable and uncontrollable events." Sir Winston Churchill, *My early life: A roving commission* (New York: Charles Scribner's Sons, 1930), p. 214.

3 Joint Publication 1-02 Department of Defense Dictionary of Military and Associated Terms, (Washington, D.C.: Government Printing Office), p. 296.

distributing and applying military means to fulfill the ends of policy," while Colin Gray describes strategy as "the threat and use of force for political reasons." A short-hand definition often used at war colleges is "relating ends, ways, and means to achieve a desired policy goal."

The next minefield is the process. Even if we think we know what we mean by "strategy," we need a way to make it. Here, good intentions intrude. In most Western political systems, strategy is created both top-down and bottom-up. In theory, political leaders come into office with a few big ideas, departments and ministries are consulted, a deliberative process follows, and decisions are taken.[4] Alternatively, an unforeseen crisis occurs, desks are cleared, very senior people huddle, rapid decisions are reached, and action follows.

Actual strategy is more opaque than these simple, clean models. Egos disrupt rational analysis. Institutional agendas trump overarching national interests. Current crises crowd out long term planning. Personal relationships dominate or shut down formal processes. Budgets constrain strategic choices. Media leaks frustrate confidentiality. Domestic politics elbows in, and re-election politics distort altogether. Real strategy-making is at best strenuous and exacting; at worst muddled, frustrating, and decidedly sub-optimal.

And strategy matters. In the domain of armed conflict (and here we are not discussing political or business or diplomatic "strategies," but strategy in its classical sense), the price of failure can be high. The extreme penalty for failed strategy can be the fall of governments, loss of territory, even the destruction of the state itself. But lesser penalties are exacted as well, in the loss of power and influence, in economic collapse or distress, in less capable and credible political and military institutions, and in a failure of national confidence and will. Strategists must ever bear in mind that in taking the state to war, victory becomes an end in itself. Even apart from the aims of the war, defeat can shatter or debilitate the

4 An example is the very methodical Afghanistan Review of the new Obama administration in early 2009.

state for years to come, possibly leading to permanent and irrevocable decline. Put another way: avoiding defeat can become the overarching aim—independent of the original strategic objective.

Military leaders work hard to overcome the frustrations and the unknowns of strategy through a deliberate planning process, a comprehensive and detailed approach to problem solving that can takes months and even years to complete. Seasoned commanders know that no plan "survives contact with the enemy"—meaning every situation is unique and will require unique solutions. But the laborious study, assessment, and analysis that goes into a detailed plan provides context, understanding, and much useful preparatory work, particularly in the logistical and administrative preparations needed to move large forces to remote locations and keep them there. Good planning provides a foundation from which to "flex" according to the situation at hand.

Political leaders usually approach strategic problems differently. Most are lawyers or business people with substantial political careers behind them. Generally, they may lack patience with military detail, may distrust strong military types, are keen to assert civilian control, and focus more on broad objectives than on the ways and means of strategy. Quite naturally, past experiences and processes that have worked well in the political or business arenas are applied to military problems with quite different results. Casual observers might think that strategy-making at the highest levels is a sophisticated, deliberative process conducted by civilian and military officials, prepared by arduous academic training informed by practical experience. All too often it isn't.

Ideally, both civilian and military leaders will forge synergistic, interactive, mutually dependent relationships. Good will and mutual respect will go far to reconcile different cultures and perspectives in the interest of teamwork and battlefield success. But the dialogue will always remain unequal. In this regard, the fashionable view that civilian leaders can, and should, intrude at will into the professional military domain is both wrong and dangerous.[5] While they may overlap, there are clear lanes that distinguish appropriate civilian and military areas of responsibility and expertise. All parties understand and accept the doctrine of ultimate civilian control. But "asserting civilian control" is the poorest excuse for bad strategy.[6]

Civilian leaders have an unquestioned right to set the aims, provide the resources, and identify the parameters that guide and bound armed conflict. They have an unchallenged right to select and remove military commanders, set strategic priorities and, when necessary, direct changes in strategy. For their part, military leaders have a right—indeed, a duty—to insist on clarity in framing strategic objectives and to object when, in their best military judgment, either the constraints applied or the resources provided preclude success.[7]

Prior to assuming office, many political leaders have little interaction with the military and with strategy itself. The lack of strategic training and practical experience cited above, if combined with a contempt for military expertise, can dislocate strategy altogether. Often in the post-war era, we see strategies advanced where the "level of ambition" outraces the resources provided, leading to protracted, costly, and open-ended ventures with decidedly unsatisfying outcomes.[8]

In contrast, military leaders are generally cautious about use of force and, if ordered to fight, argue for larger—not smaller—forces. The military preference for avoiding wars is based on a historical appreciation for how

5 Eliot Cohen is a primary exponent of this view. See his "Supreme command in the 21st century," *Joint Force Quarterly*, Summer 2002.

6 An apposite example is Secretary Rumsfeld's tinkering with military deployment orders in Afghanistan and Iraq. See Peter Shane, *Madison's nightmare: How executive power threatens American democracy*, (Chicago, IL: University of Chicago Press, 2008), p. 73.

7 This is the great lesson of H. R. McMaster's classic *Dereliction of Duty*. It is now clear that the Joint Chiefs knew, early in the Vietnam conflict, that the strategy adopted would likely fail. But individually, and as a body, they both supported and enabled it, resulting in 58,000 U.S. deaths, hundreds of thousands wounded, and a defeat that would take a generation to overcome. See H.R. McMaster, *Dereliction of Duty* (New York: Harper Collins, 1997).

8 The short massive campaigns waged in Grenada, Panama, and the Gulf War stand in vivid contrast to Vietnam, Iraq, and Afghanistan in this regard.

quickly violence gets out of hand, how devastating less-than-total victory can be for military institutions, and how painful and expensive success can be. The military preference for large forces is likely grounded in an intuitive understanding of the complexity and unpredictability of conflict.

One way to deal with these uncertainties is to overwhelm the problem with mass at the outset (Operation Desert Storm being the obvious case in point). Larger forces, though harder to manage and more costly in the short term, provide more options and greater leverage amidst uncertainty, often leading to fewer casualties and lower costs than long, open-ended conflicts. In a sense, they "smother" the friction of war and increase the chances of quick, decisive campaigns. Smaller forces, emphasizing standoff air and seapower and special operations, may seem more "transformational," but the historical record is on the side of the bigger battalions. "Transformation" has lost some at least of the glamor it enjoyed a decade ago, while more traditional approaches have grudgingly regained ground, as seen in the "surges" of Iraq and Afghanistan.[9]

Political leaders and strategists should also be mindful of *strategic culture*, that mélange of history, tradition, custom, world view, economy, sociology, political systems, and more that largely shapes how nations fight and for what causes. There may be no agreed upon American theory of war, but when based on concepts of joint and combined warfare, an "American Way of War" surely obtains mass, firepower, technology, strong popular support, and a focus on decisive battle. "Good wars" have historically followed this pattern. "Bad wars" have not. While the analogy can be taken too far, it captures central truths that should inform our strategic calculations. Strategic culture is real and powerful, whether we acknowledge it or not.

So how to square the circle to make effective strategy? Clausewitz, of course, posited that the ideal solution was to combine the statesman and the commander into one—influenced, no doubt, by the experiences

9 See H. R. McMaster, David Gray, and R. D. Hooker, Jr., "Getting Transformation Right," *Joint Force Quarterly*, 38, Summer 2005, p. 27.

of Prussia at the hands of Napoleon. Those days are long gone, never to return. The challenge today is to optimize strategy in an interagency, highly political, multinational decision setting characterized by multiple threats, declining interest in and knowledge of military affairs, financial stringency, and limited reservoirs of public support. With this in mind, the following strategic considerations might usefully be kept in mind.

Understand the Nature of the Conflict

This sounds easy but usually isn't. Both in Iraq and Afghanistan, as well as in Vietnam, the U.S. seems to have fundamentally misunderstood the nature of the conflict and to have persisted in the error for far too long. In each, intervention and initial success signaled not the end but the beginning of a long, expensive, tortuous conflict that dragged on far longer than most experts predicted. This is not Monday-morning quarterbacking. Strategists must be able to "appreciate the situation" and make concrete assessments of the problems to be solved and how to solve them. Just as importantly, they must be ready to jettison failed policies and strategies and make new ones when needed.

Consult Your Interests First and Your Principles Always

Thucydides cautioned that states typically go to war for reasons of fear, honor, or interest. That doesn't mean they always should. Interventions for moral or prestige reasons will always have a certain appeal. But there will always be ample opportunities for that. When vital interests and national values coincide (as with the first Gulf War), the prospects of strong domestic support and ultimate success are immeasurably enhanced. When they don't, expect trouble.

Don't, Unless You Have to

Colin Powell kept the following quote from Thucydides on his desk: "Of all manifestations of power, restraint impresses men most." Military adventurism can be exhilarating when viewed from a distance.[10] The sheer exercise of power for its own sake has an undeniable appeal, not often admitted by insiders. But all too often, war takes on a life of its own, and what seemed easy at the outset can become painful and difficult. Democracies, in particular, can tend towards "no win-no lose" approaches to conflict that seek to achieve grand strategic objectives with limited means, uncertain popular support, and a very low tolerance for casualties. This is not to say that only wars for survival should be fought (the Rwandan genocide comes to mind as a catastrophe that could and should have been prevented through the use of force). It is to say that as a general rule, wars of choice should be avoided, and when fought, they should be fought to quickly win with crushing force.

Political Problems are Rarely Solved with Force

Throughout history we see attempts to use military power to solve political problems. This approach overwhelmingly fails unless the adversary is crushed absolutely and his society remade. Force can help eliminate or reduce violence and set conditions to support a political settlement, a valuable and important contribution, but it cannot solve ethnic, tribal, ideological, or inherently political conflicts in and of itself. This is perhaps its most important limitation. Military power often appears to be useful because it is available and, superficially, both multipurpose and multi-capable. But it is best used to solve military problems.

Expect Bad Things to Happen in War

Clausewitz made much of the tendency of war and violence to run to extremes.[11] The famous British Admiral John Arbuthnot Fisher echoed

10 As aide de camp to the secretary of the army in 2001–2002, the author personally witnessed this phenomenon in the Pentagon. After 9/11, there was much talk among political appointees about "putting heads on sticks." The uniformed military were considerably more sober.

11 Carl von Clausewitz, (1976). *On War*, p. 77. Michael Howard and Peter Paret (Eds.). Princeton, NJ: Princeton University Press.

Clausewitz when he said "the essence of war is violence . . . moderation in war is imbecility." That goes too far, but restraint is often the first casualty in war. For soldiers, but also for statesmen, war is a struggle for survival. And the struggle for survival is inherently impatient with limits.[12] Strategists must understand and accept this basic truth. Civilians will be hurt, war crimes will occasionally happen, the press will likely not be an ally, and every mistake will be exploited by the political opposition. All these should be expected. Don't think they can always be controlled or avoided.

If You Start, Finish—Quickly

It is, or should be, a maxim of war that the longer things take, the worse they tend to get. The Weinberger and Powell Doctrines—essentially, arguments to fight only for truly important objectives with overwhelming force—were largely discredited in later years, by both political parties, and with tragic results. Because popular support is finite and rarely open-ended, democracies can ill afford long, drawn out, inconclusive conflicts. In the end, they can cost far more—in lives, treasure, political capital, and international standing. Extended conflicts provide time and space for war's natural tendency to get out of hand. Short, sharp campaigns must ever be the ideal.

In War, the Military Instrument Leads

In peacetime (defined as the absence of armed conflict), diplomacy and diplomats have the interagency "lead" as first among equals, under the ultimate control of the head of government. (It goes without saying that congressional or parliamentary support is also essential.) During times of war, the interagency lead must pass to the defense establishment, personified by its civilian head. In a sense, all wars or conflicts represent a failure of diplomacy, in that a ruling has been made that

12 See the author's "Beyond Vom Kriege: The character and conduct of modern war," *Parameters*, Summer 2005.

the state's strategic objectives cannot be realized except through force. This does not mean that diplomacy ceases or that it is unimportant. On the contrary, sound strategy demands that the type of peace we desire remains uppermost in our councils and deliberations, and that channels—often indirect—remain open even during war. But for all that, war has its own ineluctable logic, a logic that strains against the kind of modulated, nuanced, "signaling" often favored by diplomats. Even as the battle rages, they must have their say, but the louder voice should be the secretary or minister heading the defense apparatus and his Chief of Defense staff.

Don't Be Seduced by Airpower or Special Operations

When selecting strategic options, statesmen are often encouraged to choose "safer, easier, cheaper" options, relying principally on airpower and special operations forces. The prospect of lower casualties and quick wins is always seductive. Airpower is the jewel in America's strategic crown, while special operations units have undeniable appeal as versatile, high-quality forces. Neither, however, is a panacea and both, like all forms of military power, have real limitations—chiefly, an inability to control (as opposed to influence) the ground and the populations who live there. Almost always, a balanced application of military force will be needed to achieve decisive outcomes in major wars.

More Money Does Not Equal More Defense

In the U.S., at least, it can be politically dangerous to argue that more defense spending may not equate to a safer America. The combination of extraordinarily powerful defense industries, strong congressional support, and willing military leaders makes controlling defense costs a herculean task. Yet responsible strategists must confront cost and risk as necessary elements of the game. When two B2 stealth bombers cost more than the entire inventory of main battle tanks in the active army, something is badly wrong. Political dynamics and service advocacy can't be removed from the politics of defense, but statesmen and strategists should still fight for strategic balance.

Watch the Partisan Politics

Wars should always be waged to achieve political objectives, not just military ones. But statesmen should also beware of unduly politicizing armed conflict. American political debates and controversies used to "stop at the water's edge." Today, strategies may often be adopted simply because they are not what the other party advocates.[13] Democrats in Congress in 2008 strenuously opposed the Iraq surge mostly because it was a Republican idea. Republicans in 2011 fought President Obama over war powers in Libya for political, not strategic, reasons. A good rule of thumb is to set realistic, achievable political aims; get the strategy right; and then work hard to achieve consensus and bipartisan support. Policy should always drive strategy. Partisan politics is a different thing altogether.

One Crisis at a Time

Historically, senior leaders struggle to cope with simultaneous crises. By definition, every war is a crisis. Even in democracies, the circle of actual decision makers is surprisingly small. Though most won't admit it, these few cannot devote their scarcest resource—their time—to effectively managing more than one complex crises at once. This argues for a limited and conservative approach to uses of force whenever possible, or as Lincoln put it when urged to declare war on Britain during the Civil War, "one war at a time."

13 On taking office in 1992, the Clinton administration withdrew most U.S. troops from Somalia, then waged an aggressive kinetic campaign against factional leader Mohamed Farah Aidid. The result was a military and political disaster that prevented any response to the Rwandan genocide on grounds of "political" liability. See the author's "Hard day's night: A retrospective on the American Intervention in Somalia," *Joint Force Quarterly*, 54, 2009.

Keep Things Brutally Simple

In 1944, General Eisenhower was given the following directive: "you will enter the continent of Europe and . . . undertake operations aimed at the heart of Germany and the destruction of her armed forces." The mission statement in Afghanistan in 2014 was by contrast far more ambiguous and vague.

> In support of the government of the Islamic Republic of Afghanistan, ISAF conducts operations in Afghanistan to reduce the capability and will of the insurgency, support the growth in capacity and capability of the Afghan National Security Forces (ANSF), and facilitate improvements in governance and socio-economic development in order to provide a secure environment for sustainable stability that is observable to the population.[14]

If soldiers and voters are to understand and support something as serious as war, it should be explained—and explainable—in brutally simple language. If we can't, then we should go back to the drawing board.

Limit Your Level of Ambition

U.S. interventions in Korea, Vietnam, Iraq, and Afghanistan evolved into protracted, painful, debilitating and indecisive conflicts. In contrast, Grenada, Panama, and the Gulf War were quick, overwhelming successes. The difference between the two categories is apparent. A strategy with clear, achievable aims matched with ample resources usually wins. Fuzzy, overly ambitious goals supported by inadequate resources usually don't.

Get Comfortable with Bad Options

When President George W. Bush decided to send 30,000 additional "surge" troops to Iraq in late 2007, he did so against the advice of the

14 ISAF website, 2012.

uniformed military, the secretaries of state and defense, and his vice president. "Doubling down" when most thought defeat was inevitable was not popular. Few things in war are. A clearly optimal way ahead rarely presents itself. For strategists, reality can often mean choosing the least worst from a range of "bad" options. The courage to make hard decisions is necessary equipment for strategists.

Challenge Your Assumptions

Ideology can be a primary driver for use of force, and it is often buttressed with bad assumptions. In Vietnam, LBJ and his principal advisers assumed that North Vietnam was a proxy for Moscow, not a true nationalist movement. In Kosovo in 1999, both NATO and U.S. senior leaders assumed that a quick, one-week air campaign would drive Serbian forces out. In Iraq in 2003, the administration assumed that the Shia would welcome U.S. troops enthusiastically, that the Iraqi army could be disarmed without consequence, and that a makeshift transitional government could be quickly assembled to take over post-conflict responsibilities. Few of these assumptions were submitted to searching analysis, nor were contrarian views allowed to challenge the prevailing consensus (In this regard, a "Devil's Advocate" like Robert Kennedy during the Cuban Missile Crisis can be a game saver.) No strategy can overcome flawed assumptions.

Think Through 2D and 3D Order Effects

Thinking through the problem is not a strong suit of Western strategists. When Baghdad falls, what can we expect on the day after? Will massive amounts of international assistance fuel widespread corruption? Can the host nation military maintain order, or should it be disbanded? Are sanctuaries in Pakistan an unsolvable problem? In most cases, these are not unknowables. They are the bread and butter of responsible strategy-making. Senior political and military leaders must demand and enforce thorough and painstaking strategic assessments to think the problem through and beyond the initial objectives.

Tomorrow's Crisis is Not Predictable

Despite a plethora of intelligence agencies and scores of "experts" in and out of government, forecasting the next crisis is unlikely, as the Arab Spring demonstrated yet again. Pearl Harbor, the German attack in the Ardennes, the North Korean attack across the 38th parallel, the collapse of the Soviet Union, Saddam's invasion of Kuwait, 9/11, and the Arab Spring—these intelligence failures are the rule, not the exception. Broad, overarching strategies, as well as more tailored regional and theater ones, must be flexible enough to react to the unforeseen. Crisis response mechanisms must be thoroughly rehearsed. Strategists must expect surprises and be ready.

It's Always Better with Allies

Churchill loved to say, "the only thing worse than fighting with allies is fighting without them." Going it alone may be needful on very rare occasions, but in general, a lack of allies should give serious pause. Fighting with allies confers political legitimacy as well as extra troops. Often, allies can provide counsel and an outside perspective that can usefully enhance strategic decision making. Sometimes, a threat may be so real and immediate that a unilateral use of force is the only resort. Still, when long-standing allies with congruent interests balk, maybe it's time to take a deep breath and think again. Maybe they know something we don't?

Beware of Service Agendas

Each military service has its own culture, and all will fight to maximize freedom of action and access to resources. If taken too far, weighting one service risks unbalancing the strategic equilibrium that represents America's true military strength. A military that is only globally effective in one dimension, or at one gradient along the spectrum of conflict, is a military shorn of the versatility and synergism that underpins America's true military dominance. Strategists should look for service agendas and ensure they do not corrupt military planning. In general, theater and operational commanders should control the military assets and activities present in their theater under the principle of unity of command. (This is particularly true for special operations forces.) Service-specific strategic approaches should be viewed skeptically.

Carefully Count the Cost

We often hear the catch phrase "blood and treasure," but we don't always think deeply about what it means. Experienced commanders know what it means to lose soldiers in war. Statesmen should also reflect on the terrible price of victory, and the terrible penalty of defeat, when contemplating the use of force. The prospect of casualties should not deter them from making the tough decisions when they are right and necessary. The prospect of needless or unnecessary casualties—and inevitably, this will include innocent civilians—absolutely should. The financial cost of war can also be tragic. The cost of the war in Iraq, projected by the U.S. administration at $50 billion (financed, the public was told, largely by Iraqi oil revenues), approached $1 trillion—a truly colossal sum.[15] The terrible irony is not only the price in lives and dollars, but also that so many "experts" were so far off the mark. Whatever that's called, it's not strategy.

This short paper is not a strategic *tour d'horizon*, but hopes to offer practical observations and recommendations, grounded in history and a classical understanding of strategic thought. Some will find it excessively cautious or conservative. But all questions of war and peace should be approached so. Military action remains an indispensable tool in what remains an unstable and dangerous world. Yet it is one of many, often exacting a

15 In a March 29, 2012, report the U.S. General Accounting Office reported direct costs of $823 billion. Indirect costs will far exceed $1 trillion.

terrible price. Statesmen and soldiers alike are well advised to think long and deeply before sowing the wind. Sound and sober strategy-making can show us how.

3 National Security Decision Making and the NSC

Effective strategy, in peace as well as war, can only take place inside an orderly process that frames strategic problems and proposes workable solutions. Too often, the US interagency process falls short due to bureaucratic infighting, clashing personalities, a crowded policy agenda, and distracted presidents. Here the author suggests that presidential leadership and high-quality, talented practitioners—not organizational tinkering—can optimize national security performance.[1]

Early in every administration, the new president and his national security team are inundated with studies offering advice on how to organize the incoming administration for national security. Many propose sweeping changes in the size, structure, and mission of the National Security Council (NSC) staff, the fulcrum of national security decision making. However attractive superficially, organizational tinkering is unlikely to drive better performance. This paper argues that structure and process are less important than leadership and talent in national security. No duty rises higher than the president's call to defend the Constitution and the people and territory it nourishes. That duty will be tested early and often.[2]

1 This chapter is a revised version of the author's *The NSC: New choices for a new administration*, National Defense University Press, 2016.

2 A note on terminology: though the expression "NSC" is often used to denote the National

In every administration the NSC staff leads this effort, in theory if not always in practice. It has four primary roles: (1) to advise the President in the field of national security affairs; (2) to manage and coordinate the interagency process in formulating national security policy; (3) to broadly monitor policy execution; and (4) to staff the president for national security meetings, trips, and events. Many assume that the NSC staff does, or should do, much more. But it is first and foremost the President's personal national security staff.[3] Other tasks—such as generating independent, whole-of-government national security policies and strategies, or conducting detailed, daily implementation oversight—would require a much larger staff. Inevitably, it would lead to ponderous, centralized, and ultimately dysfunctional behaviors that would prevent responsive support to the president. Long range "strategic" planning is surely essential, but it more properly belongs to the interagency as a whole, vetted by the Deputies and Principals Committees and approved by the full National Security Council.

The NSC's role as a process manager is not synonymous with policy advocacy. While the national security advisor may and often will recommend a given course of action, a more critical function is ensuring that all viewpoints are heard and objectively assessed, and that important issues are framed for decision. When allowed to become an operational entity (as in Iran-Contra) or to effectively preempt the Departments of State and Defense (as in the Nixon, Ford, and Carter eras), the NSC staff has historically stumbled.[4] Properly focused and chartered, the NSC staff can

Security Council staff—that is, the body that supports the president and national security adviser—the term is properly used to mean the body chartered in the 1947 National Security Act as amended. The National Security Council's statutory members include the president, vice president, secretary of state, and secretary of defense, with the Chairman of the Joint Chiefs and Director of Central Intelligence (later the Director of National Intelligence) serving as statutory advisers. (The Secretary of Energy was added to the NSC by legislation in 2007) Other officials attend meetings of the NSC as directed by the president. See P. D. Miller, "The contemporary presidency: Organizing the National Security Council," p. 593, *Presidential Studies Quarterly*, *43*(3).

3 Amy Zegart (1999), *Flawed by design: The evolution of the CIA, JCS and NSC*, p. 79, Stanford, CA: Stanford University Press. Importantly, the NSC staff does not support the NSC per se but rather the president and NSA.

4 See Karl F. Inderfurth and Loch K. Johnson (2004), *Fateful decisions: Inside the National*

empower and facilitate an interagency process that is otherwise cumbersome. Over time, the NSC staff has become immersed in policy detail and in responding to urgent or crisis events, fed largely by a 24-hour news cycle. This in turn creates pressure for staff growth. The result is a diminished ability to conduct high level, far-seeing policy work at the appropriate strategic level. A smaller NSC staff, by definition, is unable to immerse itself in detailed policy oversight and micromanagement, a compelling argument for a leaner, more agile staff.

In this regard, the NSC staff is not a line entity, statutorily empowered to give orders in its own name. And significantly, it should not be an interagency planning headquarters. It may forward presidential guidance and direction through formal channels, or by an approved interagency body like the Principals Committee. It cannot direct or demand. However, the NSC staff and its head, the national security adviser, enjoy two distinct advantages: access to the president, and the ability to set the policy agenda in national security affairs. Used judiciously, these represent real power.

Staffing the NSC

Like any organization, the NSC succeeds based on its leadership, beginning with the national security adviser or NSA (more formally styled "Assistant to the President for National Security Affairs"). The NSA acts principally as a direct adviser to the president and as the primary manager of the interagency process.[5] The NSA must have or develop a close relationship with the President based on mutual trust and confidence, free from the intervention of other White House political agents and with direct access.[6] Historians and scholars generally agree that, among competing examples, the pragmatism, collegiality, and quiet authority of Brent Scowcroft in the administration of George H. W. Bush

Security Council, pp. 85-95, Oxford, UK: Oxford University Press.

5 Alan G. Whittaker et al. (2011, August 15), *The national security policy process: The national security council and interagency system*, p. 27, National Defense University.

6 Kevin Marsh, "The contemporary presidency: The administrator as outsider: James Jones as national security adviser," p. 831, *Presidential Studies Quarterly*, *42*(4).

represents a high standard that few others achieved.[7] (Today's interagency system of Principals and Deputies Committees and interagency working groups dates from his tenure). Though a strong and sturdy personality, Scowcroft shunned the limelight, taking his "honest broker" charter seriously and bringing clear focus, deep experience, and sound values to his high position. Scowcroft charted a middle course between egoism and desire for control on the one hand, and excessive collegiality and power-sharing on the other. Dominant personalities who seek to control outcomes—restricting cabinet officer access to the president and preventing open airing of interagency views—may well cause system failure. But weaker personalities who prize consensus above sound policy outcomes may also fail.

In selecting the national security adviser, Presidents should look for a seasoned national security professional who is able to interact with cabinet officers as an equal—without dominating them or being unduly deferential. The NSA should have a comprehensive and practical understanding of the defense and foreign policy establishments and ideally will have worked in both. (In general, academics, businessmen, and attorneys with weak backgrounds in the Pentagon and Foggy Bottom have not excelled.) A personal relationship with the president is a critical asset, but a decisive and selfless character, followed by competence and experience, are also key. A good choice will settle quickly into the role, function effectively and out of the spotlight, and last.[8] Poor selections lead to high turnover or policy drift—both painful disabilities. Some presidents have used their NSAs as public figures, deploying them on Sunday talk shows or as high-profile diplomats. Others have clearly relied on them for emotional support or as policy bludgeons. In the end, the President is free to choose

7 "The NSC system Brent Scowcroft ran for President Bush is widely held up as the gold standard." Mark Wilcox, "The National Security Council Deputies Committee: Engine of the policy process," p. 23, *Interagency Journal.*

8 The Reagan administration saw six different NSA's. This turnover contributed to any number of national security miscues and mishaps. Richard Best (2011, December 28), "The National Security Council: An organizational assessment," p. 18, *Congressional Research Service.*

what role the NSA will play. But seven decades of history illuminate the many pitfalls.

The NSA's role is most challenging when principals disagree on solutions to major problems. At these times, the NSA may elevate a contentious issue to the president for decision, and for critical issues that is both important and needful. But the president's time is scarce, and publicly overruling a cabinet officer, though occasionally necessary, is never preferred. In most cases, Principals and Deputies can and should come to consensus on policy solutions that, while perhaps not optimal from an agency/department perspective, address the overall policy objective with a feasible and practical way ahead. The ability of the national security advisor to work behind the scenes to forge that consensus represents the true center of gravity of the interagency. An effective NSA should be measured by how well these tough policy issues get solved—without constant presidential intervention.

None of this is easy. As one expert has observed:

> Our expectations of National Security Advisors are altogether unrealistic. We want them to be master administrators who advance the multilayered interagency committee process in a timely, transparent, and comprehensive fashion. But we also want them to be foreign policy and national security maestros who combine a comprehensive appreciation of the international system and security environment with a wide range of subject matter expertise across an incredible array of multifarious complex problems that enables them to discreetly offer sagacious advice when circumstances, or the President, demand it. We also insist that Advisors have an exceptionally close personal relationship with the President, essentially serving as the President's alter ego on national security.[9]

9 Christopher Lamb, "National security reform," in R. D. Hooker, Jr., (Ed.) (2017), *Charting a Course: Strategic Choices for a New Administration*, Washington, D.C.: National Defense University Press.

There is merit in this critique. The president's NSA pick is clearly critical, but we need a system that does not demand more than we can reasonably expect.

In this regard, strong calls have been made in recent years for legislation making the NSA a Senate confirmable position required to testify before Congress.[10] This would be a mistake on many levels. All presidents need and deserve confidential advisers who are free to proffer their counsel and support in confidence. Dragging senior presidential staff before the Congress can fuel highly partisan disputes with no real improvement in government performance. Through many different venues, presidents can be called out and even punished for poor performance by their NSAs.

The deputy national security adviser (DNSA) is likewise a critical appointment. Also an Assistant to the President (like the NSA), they will chair the Deputies Committee, where most policy formulation takes place, and are usually chartered to look after the day to day operations of the NSC—in short, "to make the trains run on time." (The Clinton administration also installed an NSC "Chief of Staff," while the Obama and Trump administrations combined the titles of NSC Chief of Staff and Executive Secretary.) The Deputy should have both policy and leadership experience at a high level in the national security arena. Weak government experience will limit performance in this key position. Extensive interagency experience and a collegial but decisive demeanor are prerequisites. The George W. Bush administration also created a number of DNSA's for specific areas, such as regional affairs or promoting democracy, with Deputy or Special Assistant to the President status; these ranked above NSC Senior Directors but below the "principal" DNSA. Some of these positions were continued under Presidents Obama and Trump.

Below the NSA and deputy, NSC senior directors head the dozen or so directorates which make up the NSC staff. They often hold White House "commissioned officer" rank as Assistants, Deputy Assistants, or Special

10 For example, Chairman of the House Armed Services Committee Congressman William Thornberry offered this change as an amendment to the 2017 National Defense Authorization Act.

Assistants to the President. They receive official commissions signed by the president and other privileges, such as use of the White House Mess. They head NSC offices of two to ten people and may come from government or private sector backgrounds. They are typically more experienced, perhaps more partisan, and will often stay for longer tours than detailed directors. Senior directors regularly attend Deputies Committees (DC) as direct participants, and Principals Committees (PC) in support of the NSA. They may also chair senior interagency working groups (such as the Iraq Steering Group in the Bush '43 administration) below the deputies level. Depending on their responsibilities, Senior Directors may have significant exposure to the president.

Most rank and file staff members are detailed from the intelligence community, state department, the military, or other government agencies. They are paid by their home agencies and are normally assigned for one or two years. Additionally, a few staff members may come from the private sector and are paid out of the NSC operating budget. Styled as "Directors," they are the action officers who perform most of the NSC's day to day staff functions. Each must be qualified to hold extremely sensitive clearances.

NSC directors are assigned large portfolios and are expected to work without detailed supervision in leading the interagency in their assigned areas. Many have significant government experience, outstanding records, and impressive graduate educations. As most are career government employees, they provide a somewhat less partisan environment and "feel" to the NSC in comparison to other White House offices. Directors lead interagency working groups and will often attend Deputies Committees as backbenchers and subject matter experts. As a rule, directors have limited exposure to the president.

In general, the president is best served when NSC staff members are carefully screened for their intelligence; academic and professional qualifications; experience; and willingness to serve out of the limelight. The ability to work quickly and to a high standard under stress, to fit smoothly into an interagency team setting, and to operate comfortably around very senior government officials is needed. Stamina and imperturbability are highly prized.[11] In this regard, agency reluctance to offer up their best talent should be met directly and head on. The president needs and deserves talent and quality and should get it.

More often than not (although there are occasional exceptions), hiring young, inexperienced, unseasoned NSC staffers—even with impressive academic credentials or connections to prominent figures—will backfire. Excessive ambition, inexperience, lack of credibility, personal agendas, and divided loyalties can only distract and disrupt complex and highly sensitive NSC machinery. The NSC staff is a place for grownups, committed to serving the nation and the president. Only careful vetting and objective criteria will ensure an NSC staff that is up to the task.

The Interagency

The interagency process has in general stood the test of time, but is slow, often unfocused, easy to obstruct, and obsessed with consensus. All too often, participants will defend agency positions in the certain knowledge that good policy solutions are thereby prevented. A certain amount of inefficiency is accepted in the interests of comity and the widest possible "buy-in." Nevertheless, a "lowest common denominator" solution—what former Secretary of State Dean Acheson described as "agreement by exhaustion"—is seldom optimal.[12]

NSC staff members have often, though not always, chaired interagency working groups.[13] Standing working groups are usually established by presidential directive early in an administration, with ad hoc groups set up by the Deputies Committee for more specific issues.[14] These groups

11 Whittaker, p. 41.

12 Cited in Inderfurth, p. 43.

13 Nomenclature changes from one administration to the next; past terminology includes "Interdepartmental Groups," "Interagency Groups," "Interagency Working Groups," "Policy Coordinating Committees," and "Interagency Policy Committees."

14 See for example National Security Presidential Directive 1, "Organization of the National

are essential cogs in the policymaking process and, when well-run, will generate most effective policy solutions. Sometimes co-chaired by a representative from the designated "lead" department or agency, depending on the issue, they may meet weekly or more often. NSC chairs work hard to ensure that group membership remains stable; that policy problems are well defined and scoped; that all useful views are considered; and that working papers forwarded to deputies meet high standards for completeness, accuracy, and brevity.

However, success is often elusive. One reason is that the interagency is actually many interagencies. The Obama administration's Counter-ISIL campaign is an apposite example. Multiple working groups focused on Iran, Iraq, Syria, Turkey, Russia, Intelligence, and Counter-Terrorism collided in a kaleidoscopic mélange of competing interests and priorities, leaving deputies to sort out a bewildering set of opinions and recommendations. Effective integration and synchronization of multiple "lines of effort" or LoE's foundered on a particularly remarkable example of bureaucratic disarray.[15] A more effective approach would be to constitute a Steering Group, composed of the LoE leads and led by an empowered Presidential Special Envoy, to coordinate LoE activities, track progress against approved performance metrics, and report regularly to Deputies and Principals. Key decisions would be framed for decision by the NSC staff.

In theory, most of the heavy lifting should be done in the Deputies Committee, where the bulk of policy decisions are expected to be made. As envisioned, DCs should be staffed by department and agency number twos or number threes; meet at regular intervals to consider working

Security Council System," dated February 13, 2001, and signed by President George W. Bush, available at https://fas.org/irp/offdocs/nspd/nspd-1.htm

15 "Inadequate unity of effort plagues every level of this war, and it will cripple the coalition unless it is remedied. At the highest level, no single synchronizer of US government efforts has been named. US government departments or agencies are designated the leads for one of the nine lines of effort, but there is no daily orchestration of the campaign in a whole-of-government or whole-of-coalition sense." Linda Robinson (2015, August), "An Assessment of the Counter-ISIL Campaign," p. 4, The RAND Corporation.

group recommendations; and where possible, make policy decisions without further reference to higher levels. Broad oversight of policy implementation is inherent in this charter.[16] For the most important policy issues, Deputies are expected to shape issues for presidential decision by refining working group products and clearly defining points of convergence and dissensus.

In practice, the effectiveness of the Deputies Committee has been hampered over time by too frequent meetings addressing lower priority issues with both an expansion and a dilution of committee membership. At the height of the Iraq War in 2007, DCs on Iraq were held weekly or even more often. The senior Pentagon officials present were often the Deputy Assistant Secretary of Defense for the Middle East and the two-star Vice Director for Strategy, Plans, and Policy—far below the Deputy or Under Secretary of Defense and four-star Vice Chairman of the Joint Chiefs that was originally envisioned. Over time, offices and agencies represented in the Deputies Committee have ballooned.[17] These trends replicate, to a lesser extent, the "multiple interagencies" problem faced at the working group level, with DCs composed of sometimes different personalities meeting more and more often to consider more and more issues. "True" Deputies simply lack the time to devote to endless sessions in the White House Situation Room, pondering less important issues, when their primary job is to run their departments and agencies. As a consequence, modern DCs are somewhat less able to focus on broad policy development and strategic guidance than before. Driven "into the weeds," the Deputies Committee today lacks the influence and impact of former times.

16 The origin of the Deputies Committee can be found in a supplement to President George H. W. Bush's National Security Directive 1, issued on October 25, 1989, and authored by National Security Adviser Lieutenant General Brent Scowcroft.

17 Originally, representation in the Deputies and Principals Committees was restricted to those departments and agencies represented on the NSC by statute, with others attending by invitation based on the agenda. Over time, for reasons of influence and prestige, occasional attendance hardened into more or less permanent representation for many offices. Modern DCs and PCs may see multiple White House offices represented, for example, a practice uncommon in the late 1980s.

The composition of NSCs and PCs is more stable but has also expanded over time (the routine addition of the ambassador to the United Nations is an example). At this level the pathologies described above are more muted. Principals surely meet more often than they once did, a feature of the 24-hour news cycle and modern information flows, but as before, all national security issues that matter most come before them. As the most senior executives in government, their time is precious and their responsibilities many. Meetings of the National Security Council and Principals Committee are most effective when subordinate committees have done their work well; when Principals are asked to adjudicate those decisions that only they can decide; when meetings are executed crisply and efficiently to preserve senior leader time; and when the president is given the information they need to make effective decisions in a timely manner. The NSC staff can play a central role in managing all of these requirements to best effect.

In some administrations, interagency groups (even at high levels) were chaired by departments and agencies. This should be avoided.[18] As honest brokers and process managers, NSC representatives should chair all levels, perhaps with designated lead agencies as co-chairs. In this way, the president's policies can be consistently and uniformly applied and enforced, and narrow departmental agendas prevented from dominating or undermining the policy process.

A common refrain these days is that "the interagency is broken." Particularly in academic circles, the NSC staff is criticized for a perceived inability to properly lead the interagency to timely and effective policy outcomes. Critics have called for a stronger, larger, restructured, or more empowered NSC staff to break through the bureaucracy and overcome departmental resistance and foot dragging.[19] These views

18 This was a key finding of the Tower Commission, appointed to look into the Iran-Contra scandal and recommend improvements to NSC operations. Best, p. 19.

19 See Jack A. LeCuyer (2012, December), "A national security staff for the 21st century," US Army War College, *Strategic Studies Institute Monograph*.

reflect a fundamental misunderstanding of how the interagency actually works.

In most instances, frustration with "the interagency" is actually frustration over an inability to achieve a specific desired policy result. The NSC staff can be fairly criticized if it fails to "tee up" pressing policy issues for Deputies or Principals consideration; if it fails to ensure that realistic courses of action are brought forward; if it fails to provoke objective discussion of the advantages and disadvantages of each; or if it fails to follow up on overarching policy execution by holding agencies responsible for performance. But lack of support by one or more key principals—the most common reason a proposed policy fails—is not a process flaw. Rather it is government in action. When the NSC staff functions properly in its role as process manager, concurring and dissenting opinions are passed up quickly, honestly, and objectively to an appropriate level for decision. Issues are not routinely returned to committee, rescheduled, deferred, or ignored. Appropriate senior leaders attend on time, pre-briefed, and ready to discuss and vote; in lieu of juniors are not permitted because they do not have the standing to represent agency positions anyway. Ad hoc decision groups that exclude key players and keep no written records are banned. Intentional leaks of classified information, intended to influence policy outcomes, are not tolerated. All this takes firm leadership that drives toward effective policy without alienating key interagency leaders.

Frustration with interagency performance can lead to calls for a stronger NSC staff. Investing the NSC with policy dominance—that is, the power to determine policy outcomes regardless of departmental views—carries weighty risks. It will invariably "operationalize" the White House, with potentially harmful results. It deprives the president of a free flow of valuable agency expertise and perspective. It can lead to a "slash and burn" approach to policy that destroys interagency cooperation and encourages bureaucratic insurgency.

A word about presidential special envoys may be in order here. These have proliferated in recent administrations, probably because the president and senior White House officials feel better able to control them as "direct reports." Their appointments also signal particular presidential attention and emphasis.[20] Still, special envoys typically lack staff, funding and terms of reference to enable them to overcome or challenge cabinet officers and the departments they manage. On the whole, and with a few exceptions, their performance has not been impressive. Much like convening a presidential task force to "study" a particular issue, special envoys may generate light but not much heat. Using them as "work arounds" is not likely to materially improve the performance of the interagency process.

If the problem isn't the organizational process, what is it? In a word, it is "execution." While the inherent weaknesses of government by committee can't be completely overcome, the interagency process can be made to function better. The solution will not be found by rearranging the deck chairs.

The traditional system of interagency working groups, Deputies and Principals Committees is sound. What is unsound is the common practice of putting consensus above performance. In the final analysis, the interagency can function effectively when all know that a seat at the table confers a vote, but not a veto. Anything else leads to paralysis. Here, the role of the NSA is critical; here, they must be supported by the president. When one vote can stop a major, badly needed policy adjustment, the system breaks down. Presidential leadership, judiciously brought into play when most needed, is the key.

Size and Structure

Time has seen progressive growth in the NSC staff from less than twenty in the Eisenhower administration to more than 400 under

20 David Auerswald (2011), "The evolution of the NSC process," p. 48, in *The National Security Enterprise*, Roger George and Harvey Rishikof (Eds.), Washington, D.C.: Georgetown University Press.

Obama. The Bush '43 NSC staff had approximately 260 personnel assigned (including about spell out administrative and support staff), organized into 20 offices. Regional offices (East Asia, South and Central Asia, Western Hemisphere, and Europe, which included Russia) were grouped under one DNSA for Regional Affairs, except for the Office for Iraq and Afghanistan, under its own DNSA. (Near East and North Africa were also split out, under the DNSA for Democracy, Human Rights, and International Organizations). Functional offices included Strategic Planning, Defense, Combating Terrorism, Intelligence, Counter-Proliferation, International Economics, Legal, Communications, and Legislative. Support offices were the Executive Secretary, Systems, Records, Administration, and the White House Situation Room. A separate Homeland Security Council was also established following 9/11 (later merged back into the NSC staff under Obama). Under President Obama, the NSC staff (for a time re-labeled as the "National Security Staff" or NSS) reorganized its regional groupings; added cyber, climate change, and development directorates; and assumed a more pronounced and intrusive oversight role as it expanded in size.[21] In the Trump administration, APNSA H. R. McMaster oversaw a substantial reduction during his tenure to under 200 policy experts, though with an expanded Strategic Planning directorate.

An incoming administration may look at some restructuring of regional or functional responsibilities, but a radical reorganization will likely do more harm than good.[22] In some administrations, offices were created with cross-cutting responsibilities that invited conflict and competition. (The Office of Global and Multilateral Affairs in the Clinton administration is an example.) Without a very limited and clearly defined

21 "The National Security Council has increasingly micromanaged military operations and centralized decision making within the staff of the National Security Council." *The Congressional Record*, Deliberations of the Senate Armed Services Committee, May 17, 2016 (Washington, D.C.: US Government Printing Office), p. H2677.

22 See Robert Worley (2009), *The National Security Council: Recommendations for the new president*, p. 7, Johns Hopkins University Press.

charter, such offices will tend to stray across the policy landscape, often leading to in-fighting and intramural clashes.

Historically, the NSC staff has been quite small compared to the staffs of major agencies and departments, for good reasons. Though limited in its ability to exercise detailed planning and oversight functions, the NSC staff should be agile, outcome oriented, and appropriately focused on support to the president and their policies. Growing the staff to replicate functions found elsewhere is a standard, even formulaic recommendation by think tanks and academics. But a lean, responsive, high-performing NSC staff has stood the test of time in both Republican and Democratic administrations. Creating more bureaucracy will not improve efficiency.

Some appreciation of the role of the Office of the Vice President (OVP) is essential to understanding NSC staff functions. As a statutory member of the National Security Council the vice president is of course a central figure. Beginning with the Clinton administration, the vice president's role in national security affairs has become steadily more pronounced (although Vice President Pence has not been as active in foreign affairs as his recent predecessors). OVP is represented in all DCs and PCs and the vice president's national security adviser and national security staff participate in all meaningful deliberations. Depending on the issue and on key personalities and their relationships with other principals, OVP can play a constructive or obstructive role in national security deliberations. Importantly, the vice president enjoys direct access to the president and may often conduct closed door policy deliberations without other principals in the room. Vice Presidents Cheney and Biden in particular have exercised this influence, though Vice President Pence's role has been more muted. What seems clear is that OVP's position, power, and sway inside the NSC system are not likely to diminish. Ideally, however, OVP should not become a rival or alternate NSC staff with its own policy agenda, separate from the president's.

Rx for National Security

If this general thesis—that leadership and performance are more important than structure and process—is correct, then the president's selection of cabinet officers and senior staff are supremely important. Many variables come into play. The president may have political debts that must be paid. Geographic, ethnic, and gender diversity must be considered. Close personal relationships may enter the equation. But above all, character and competence must take precedence. Candidates who were outstandingly successful academics, lawyers, businessmen, or legislators may lack deep experience in the Executive branch and flounder when named to head large executive departments.[23]

How well the president's national security team interacts with each other is just as important. Even the best NSC staff cannot overcome fratricidal strife between cabinet officers or senior White House officials. If allowed, these may often seek to leverage access to the president to bypass or short circuit the established interagency process. Recent history is rife with examples of White House and cabinet personalities who could not or would not cooperate, greatly impairing performance.[24] Even highly qualified, exceptional leaders can be poor choices if they are unable to serve as loyal and collegial teammates, or unwilling to place national interests ahead of departmental ones. In many cases, these proclivities can be known well in advance and they should be considered in all high level personnel selections. Few presidential decisions will matter more than who sits across the table at NSC meetings.

23 There are many examples. Louis Johnson, Charles Wilson, Robert McNamara, Les Aspin, Anthony Lake, and Chuck Hagel could be cited, among others. At the other end of the spectrum, Colin Powell, Robert Gates, Leon Panetta, and, of course, Brent Scowcroft stand out as deeply experienced and effective NSC leaders and practitioners.

24 The rivalry between Alexander Haig and Casper Weinberger in the first Reagan administration is a clear case. A more recent example is the role played by Secretary of Defense Donald Rumsfeld and Vice President Dick Cheney in the Bush '43 administration, which effectively sidelined the NSC staff and more broadly the interagency process. Rumsfeld's rivalry with General Colin Powell has been well documented and largely blunted the effectiveness of Powell, otherwise considered an exceptional Secretary of State. Worley, p. 31

How well the NSC staff and the interagency as a whole performs will be crucial in determining any administration's success in keeping America safe. It is no place for oversized egos, partisan bomb throwers, ambitious juniors, overbearing seniors or policy "tourists" who lack the interest and persistence to work key issues across months and even years. Here, competence, character, collegiality, and selfless service are the keys to the kingdom. The specifics of internal structure are not particularly important. Talent and leadership, as well as a proper appreciation for roles and responsibilities, are. It will always be the president's prerogative to mold the NSC staff and make it fit for purpose. In the end, it is the president who determines the kind of process they will have.[25] But we have plenty to go on in assessing "what right looks like." The stakes are high: nothing less than the safety and security of the American people.

25 See Kori Schake and Will Wesler (2017, January), "Process makes perfect: Best Practices in the art of national security policymaking," *Center for American Progress*. Available online at https://www.americanprogress.org/issues/security/reports/2017/01/05/295673/process-makes-perfect/

The Character and Conduct of Modern War

"You may not be interested in war . . . but war is interested in you."
—Leon Trotsky

Has the nature of war fundamentally changed with the advent of new technologies and non-state actors? Many now argue that the Clausewitzian framework has been superseded and replaced by a "new" understanding of war and armed conflict. Here, the author argues for the continued relevance of classical modes of thought—married with an appreciation for the potential of new technologies to alter the conduct, but not the fundamental nature, of war.[1]

It is the tragedy of history that man cannot free himself from war. Indeed, far more than by the development of art or literature or trade or political institutions, the history of man has been determined by the wars he has fought. Time and again, advanced and cultured societies have been laid low by more primitive enemies with superior military institutions and a stronger will to fight. The end of the Cold War, the rise of globalization, the spread of democracy, and the advent of a new millennium raised hopes that mankind might move beyond the catastrophic wars that shaped the twentieth century. Those hopes were dashed by Somalia and Rwanda and Bosnia, by the Sudan and the Congo and Kosovo, by Chechnya and

1 This essay appeared as "Beyond Vom Kriege: The character and conduct of modern war," *Parameters*, Summer 2005.

Afghanistan and Iraq. Understanding war, not as we would like it but as it is, remains the central question of international politics. And for the most primal of reasons: War isn't going anywhere, and war matters.

Political and military leaders are notoriously averse to theory, but if there is a theorist about war who matters, it remains Carl von Clausewitz, whose *Vom Kriege* ("On War") has shaped Western views about war since the middle of the nineteenth century. It goes too far to declare, as John Keegan has, that Clausewitz "influenced every statesman and soldier interested in war for the past 100 years;" most never actually read or understood him. Yet Clausewitz endures, not because he is universally understood or accepted but because he is so often right about first principles.[2] Much of what he wrote about the conduct of war in the pre-industrial era, about marches and magazines and the "war of posts," fits best with his own time. But his insights about the nature of war itself remain uniquely and enduringly prescient.

Clausewitz described war as "nothing more than a duel on a larger scale . . . an act of force to compel the enemy to do our will."[3] Today, "war" is used to mean very different things in very different contexts, from the war on poverty to the war on drugs to the war on terrorism. Because it evokes a call to action and stimulates national resolve, "war" is perhaps the most used and abused word in the political lexicon. What does it mean precisely?

War is surely both a duel and an act of force, but it is perhaps best described as *armed conflict between states*. While not inconsistent with Clausewitz, this usage lends both simplicity and clarity to often-muddied waters. Thus defined, war can be distinguished from raids, rescue operations, peacekeeping missions, counter-drug and anti-terror operations, military occupations, shows of force, and a host of other activities which

2 Cited in Christopher Bassford, "John Keegan and the grand tradition of trashing Clausewitz," *War and History, 1* (November 1994), p. 13.

3 Carl von Clausewitz, (1976). *On War*, p. 75. Michael Howard and Peter Paret (Eds.). Princeton, NJ: Princeton University Press.

involve the use of military forces. Implicit in this usage is reciprocity; an unanswered, one-time cruise missile attack is a military operation and a use of force, but hardly a war. However ineffectually, however great the mismatch, both sides must participate in the "duel" for war to exist.

Nor does official sanction particularly matter. Whether formally declared or not, war is war. Nowadays, even advanced states routinely forego the diplomatic niceties, though all seek and welcome the imprimatur of international support and recognition when they can get it.

Here, "armed conflict" means fighting—not a show of force or the threat of invasion, but actual combat. The difference is important because the many gradations of the use of "forces" are distinct from the use of "force." Fundamentally, war itself is not about deterrence or dissuasion, although the capability and the will to wage it may be. As Bedford Forrest so pungently put it, "War means fighting. And fighting means killing." The distinction is crucial. The chance of stumbling into war is too great. All too often, statesmen have used the threat of war as a tool of policy—only to be astounded when it fails and war erupts.

If war is armed conflict between states, what is its purpose? *The purpose of war is to impose the will of one state on another by force.* Ideally, wars are waged for some definable, rational purpose; as T. R. Fehrenbach explained, "The object of warfare is to dominate a portion of the earth, with its peoples, for causes either just or unjust. It is not to destroy the land and people, unless you have gone wholly mad."[4]

But not always. War also can be inchoate and incoherent, its object not far removed from insensate mayhem. Sometimes, states do go mad. It may be wisdom to insist, as Clausewitz so famously did, that war conform to its political objective.[5] It would be foolish to think that it always does.

4 See T. R. Fehrenbach, *This Kind of War* (New York: Brassey's, 1994).

5 Much ink has been spilled over what exactly Clausewitz said, or meant to say, on this point. The author inclines to the side of those who believe the most famous Clausewitzian dictum ("war is a continuation of politics by other means") reflects the view that resorting to war should be a rational decision undertaken in pursuit of rational ends. Here the linkage between political ends and military means is apparent. This is not to say, however, that it always is, or that Clausewitz believed so. See Bassford, p. 6.

The term "state" also deserves precise definition. Political scientists often attach stringent conditions to statehood, but a state can be described accurately as any political entity which controls territory and population and can effectively wield power relative to its neighbors. It may be vast, like the Democratic Republic of Congo, or tiny, like Chechnya or Abkhazia. It may or may not be internationally recognized or conventionally organized. It may be ethnically homogeneous like Sweden or a tribal mosaic like Iraq. The form of government is not particularly important. What matters is the ability to exercise control internally and maintain it when challenged.

States so defined may rise and implode. They may be little more than criminal syndicates thinly disguised, like Transnistria, or patch-works of rival clans, like Somalia, or entities tortured by irreconcilable differences, like Sudan. Whether stable or failing, states matter because, among other things, they provide havens for international terrorism and transit points for the flow of arms. While the West can conduct military operations against transnational threats, as we have seen in Yemen and the Philippines, it may take more to destroy the protected enclaves that a functioning regime can offer when the financial or ideological price is right. It may take a war.

Critics have strenuously objected that the Clausewitzian thesis ignores the grave threats posed by international terrorism and other transnational actors.[6] These are indisputably and powerfully real. But one does not wage war in the pure sense against shadowy cells dispersed among many different sovereign states, some of whom are close allies and others of whom may not even be aware of the terrorists in their midst. The war in Afghanistan meets our definition because the Taliban controlled territory and population and exercised the practical functions of statehood. Except in the purely local sense, al-Qaeda does not.

This is not to say that al-Qaeda or Hezbollah or Hamas are not exceedingly dangerous. But the means used to combat terrorism, or

6 See Martin van Creveld, *The Transformation of War* (New York: Free Press, 1991).

narco-traffickers, to cite another example, lie primarily in the intelligence, law enforcement, public diplomacy, and information-sharing arenas—and only secondarily in the military sphere. This is an important point. *States are not waging war when armed force is not the primary agent.* Used imprecisely, "war" assumes rhetorical importance as a way to mobilize popular support, express seriousness of intention, and prepare the citizens for sacrifices. But the state-directed use of armed force is not the thrust of the campaign against international terrorism; it plays only a supporting role.

The vocabulary of war is important because so much is done in its name. Perversely, Clausewitz is often condemned to irrelevance by those who first redefine war and then castigate him for not describing it "accurately."[7] War as understood in the classical sense remains consistent with Clausewitz's most famous aphorism: that war is simply the continuation of political activity by other means. Explicit in Clausewitz's formulation is the notion that, because of its unpredictability and tendency toward extremes, war must be subordinated to a rational purpose and clearly defined: "The first, the supreme, most far-reaching act of judgment that the statesman and the commander have to make is to establish the kind of war on which they are embarking."[8] And while military force is only one weapon among many, with diplomatic, economic, political, informational, and even soft power instruments of statecraft available, it is by far the first among equals in wartime. Even more powerful than the impact of death on societies is the impact of ideas for which people are willing to die. Those ideas find their ultimate expression in the organized violence of states.

Today, no power on earth can compete with the United States and its allies in major conventional war, and few seem inclined to try. But war itself is flourishing, its essential nature unchanged. In northern and sub-Saharan Africa, in Central Asia and the former U.S.S.R., in Kashmir

7 See Alvin and Heidi Toffler, *War and Anti-War* (Boston: Little, Brown, 1993).

8 Clausewitz, p. 88.

and Tibet, and above all, in the Middle East, war is a growth industry. Fueled by many things, but above all by religion and economic disenfranchisement, war attracts desperate and disillusioned youth into a culture of violence. All too often, as a tool for concentrating political power in the hands of the few, rearranging the political landscape, and redirecting challenges to authority toward real or imagined enemies, war works.

The Character of War

Given the dramatic changes sweeping the globe in virtually every field, the temptation to think about war as something altogether different from before is overpowering. Indeed, advocates of military transformation in the United States assert that technology has redefined war altogether. Nothing could be more mistaken. While the methods used to wage war are constantly evolving, the nature and character of war remain deeply and unchangeably rooted in the nature of man.

Clausewitz wrote, "If war is an act of force, the emotions cannot fail to be involved."[9] The emotional or passionate side of war receives scant attention from modern theorists and policymakers, though it permeates state-to-state conflict at every level. It is easy to imagine the fear and rage and grief of the combatants, harder to see it in the cool press briefings of the leaders who make war and the often-mute suffering of the populations who must endure and support it. Yet it is ever-present. Clausewitz saw clearly that war has a nature all its own, a nature that left to itself must run to extremes. This tendency of war to run away with itself—to leap its banks and escape the original purpose of the conflict—recurs over and over in history, pressing hard against the rational courses of policy and strategy. Where does it come from?

War is much more than strategy and policy, because it is visceral and personal. Even when the existence of the state is not at risk, war in its purest form is a struggle for personal or political survival, a contest

9 Ibid., p. 76.

for the highest stakes played out directly by its participants and indirectly by the people and their leaders. Its victories and defeats, joys and sorrows, highs and depressions are expressed fundamentally through a collective sense of exhilaration or despair. For the combatants, war means the prospect of death or wounds and a loss of friends and comrades that is scarcely less tragic. But society is an intimate participant too, through the bulletins and statements of political leaders, through the lens of an omnipresent media, and in the homes of the families and the communities where they live. Here, the safe return or death in action of a loved one, magnified thousands of times, resonates powerfully and far afield.

Depending on the state's success in building popular support for war, a reservoir of endurance to losses and defeats can exist. But it is finite, its depth a measure of the public's support for the causes engaged. When it is exhausted, the government itself faces grave political risks. For this reason, if for no other, war is the ultimate gamble. For soldiers and premiers alike, war is about survival. And the struggle for survival is inherently impatient with limits.

In this stressful and highly charged environment, violence has a cascading effect as the frustrations and frictions of the battlefield encourage ever-increasing uses of force. Restraint and moderation are often the first casualties.

The mounting toll of military and civilian casualties and the images of war, seen firsthand or worldwide on CNN, beget a traumatized population and an increasingly exasperated and desensitized military. Except in very short conflicts, mounting impatience soon permeates the conduct of war, enhancing and emphasizing its inherent emotional component. War's ebb and flow may lead to changes in its aims and objectives in midcourse, either from the thrill that accompanies success or the dismay and even panic that follows defeat or stalemate. In either case, the rational and sober conduct of the war is constantly challenged and influenced by passionate and elemental currents closely related to the character of war itself. The ineluctable nature of war is summed up in the words of the German general in Russia who said, "We are like a man who has seized a wolf by the ears, and dares not let go."[10]

The passion and emotion generated by war unquestionably account for its durability and its tendency to spawn new and more vengeful conflicts afterward. Wars are difficult to win conclusively. The wars of Napoleon led to a reformed Prussian army and a revived military state that, within the same lifetime, created modern Germany and destroyed the French Second Empire. France smoldered for decades over the loss of Alsace-Lorraine ("never speak of it, never forget") and leapt eagerly into the fray in 1914. The destruction of Wilhelmine Germany and the shame of Versailles birthed National Socialism and the Second World War, from which emerged the bitter rivalry between Russia and the United States and its peripheral wars in Korea and Indochina. Today, America is at war with many of the same *mujahideen* it supported against the Soviets in Afghanistan, that Cold War spinoff of the 1980s. And on, and on. Powerful enmities are transmitted through the generations with fearful force, as though the Glorious Revolution of 1688 or the Battle of Kosovo Polje were current events and not ancient feuds.[11]

As Clausewitz noted, it is just this tendency which gives war its own trajectory, its inherent anti-deterministic and nonlinear character. The first-order effects of armed conflict between states may be apparent—the military defeat of one side or the other, an exchange of territory, the fall of a regime, or a shift in the local or international political equilibrium. But the second- and third-order effects are never as easy to predict, and may be profound in their unintended consequences. Even victory is often not the end. National populations, and the populist leaders who exploit them, do not easily forget or forgive. Taking the state to war is always a gamble,

10 Attributed to Major General F. W. von Mellinthin, Chief of Staff of the German 48th Panzer Corps in Russia in World War II.

11 "In war the result is never final." Clausewitz, p. 80.

regardless of the military balance of forces. Invariably, war will have its way. As Churchill put it:

> Never, never, never believe any war will be smooth and easy, or that anyone who embarks on that strange voyage can measure the tides and hurricanes he will encounter. The Statesman who yields to war fever must realize that once the signal is given, he is no longer the master of policy but the slave of unforeseeable and uncontrollable events. Antiquated War Offices, weak, incompetent or arrogant commanders, untrustworthy allies, hostile neutrals, malignant Fortune, ugly surprises, awful miscalculations all take their seat at the Council Board on the morrow of a declaration of war. Always remember, however sure you are that you can easily win, that there would not be a war if the other man did not think he also had a chance.[12]

Despite a terrifying increase in its scope, scale, and lethality, war persists as a political genre, first because it mobilizes and unifies the state behind its leaders as nothing else can, and second because states so often persuade themselves that they can win. Rarely do states accept battle with no hope of victory. Even the Melians expected succor from Sparta.

Intuitively building on war's nonlinear character, Clausewitz advanced his famous "trinity" as one way to describe the contending forces which affect the course of war. Often summarized as "the people, the army, and the state," the Clausewitzian trinity is actually more subtle and penetrating. He saw the emotional, inconstant force of the masses; the role of chance and probability experienced on the battlefield by the military; and the state's attempts to subordinate war's tidal forces to rational policy as a dynamic and interactive process.[13] A keen student of science, Clausewitz likened this interplay of forces to an object suspended between three magnets. Although subjected to like, measurable

12 Winston Churchill, *My early life: A roving commission* (London: Thornton Butterworth, 1930).

13 Clausewitz, p. 89.

forces, the object reacts erratically and in ways which cannot be replicated even under identical conditions—an apt analogy for war that, for all our modernity, holds true.[14]

Over the millennia, man's practical experience of war, of its horrors and excesses, has brought forth all manner of international legal codes designed to limit its extent and effects. Augustine's concepts of just and unjust war and the attempts of Grotius to regulate its conduct in law have powerfully influenced Western thought. But in the end, states most often interpret justice in light of their interest, giving the use of force an enduring place among the tools of statecraft.

Understanding war in its true form is crucially important because otherwise war can become an instrument for resolving all manner of political disputes—an exceedingly dangerous state of affairs. Especially for powerful states, whose military dominance suggests "easy" solutions for intractable problems, war cuts through the tortured legalisms of international institutions, shortcuts leaky economic embargoes, and truncates difficult and frustrating diplomacy. Power and impatience are a seductive but deadly combination, best controlled by thoroughly comprehending war as it really is. War is sometimes the right, the true, and the wisest course. Sometimes the attacked party is given no choice at all, except whether or not to resist. But a full understanding of war's tendency toward extremes of violence and its unpredictable outcomes militates against an early or easy recourse to force except under conditions of great risk.

All of this may frustrate those who believe in more expansive or less limiting definitions. Ongoing military operations in Iraq, for instance, or the "Global War on Terrorism," are proof to many that war has slipped the bonds of state-to-state conflict. In Iraq or Afghanistan, however, military force may be used to provide a secure environment, but in the current phase of "stability" operations, armed conflict (in military parlance) is not the "main effort." Public diplomacy, intelligence sharing, economic

14 See Alan D. Beyerchen, "Clausewitz, nonlinearity and the unpredictability of war," *International Security, 17* (Winter 1992), pp. 59–90.

assistance, national and international law enforcement, and many other tools are as important, or more important, than armed force in these and like instances.

The Conduct of Modern War

If Clausewitz's reflections on the character of war remain valid, his observations on the conduct of war are less apt. Few things change more rapidly than the conduct of war, rooted in the intersection between technology and the political, economic, and military institutions of the state. That trend is accelerating at a fantastic rate.

Beginning with the industrial revolution, the technology of war began to change exponentially rather than incrementally, outstripping tactics and strategy, doctrine and organization. In the American Civil War, neither side ever really grasped the impact of new technology on old ways of fighting. Fifty years later, the same could be said about the Great War. Because technology evolves so quickly, the weapons of war often outrun its methods and modalities. In general, technology has increased the distance at which man kills, enhanced the lethality of his weapons, and reduced the time needed to train him for war. For advanced, wealthy states, cutting-edge technology is accelerating trends toward smaller, more professional, and more expensive militaries who are oriented on precision weaponry and networked sensors.

As crucial as technology can be to war, other factors can and do play decisive roles. At least from the time of the Punic Wars to the time of Constantine the Great, a span of some 500 years, the Roman army bestrode the military scene and proved by far the most important factor in the growth and stability of the Roman empire. Its greatness was based not on better weapons but on its superior military institutions, expressed in careful training, organization, and discipline. These "human" factors often overshadowed technology in the centuries that followed. Although the Mongols possessed nothing like the heavily armored horsemen of

Europe, and they did not grossly outnumber their opponents as usually assumed, they created in a short time perhaps the largest empire in history, stretching from the Sea of Japan to the gates of Vienna. The victories of Gustavus Adolphus, Cromwell's New Model Army, the Prussian army of Frederick the Great, and the armies of Napoleon were not in the main the results of technological overmatch. In the modern era, the Germans, Chinese, North Vietnamese, and Israelis all achieved outstanding battlefield successes against opponents armed with equal or superior technology. Future wars may well see technology playing an even larger role, but other dimensions will still play an important part in what remains, essentially, a contest of wills played out by thinking and adaptive opponents.

Modern war, at least as practiced in the West, trades on American and European technology and wealth, not on manpower and ideology. Western militaries are typically small, professional organizations officered by the middle class and filled by working-class volunteers. Their wars are universally "out of area"—that is, not fought in direct defense of national borders—which places a premium on short, sharp campaigns won with relatively few casualties. Although land forces remain indispensable, whenever possible Western militaries fight at a distance using standoff precision weapons, whose accuracy and lethality make it difficult or impossible for less-sophisticated adversaries to fight conventionally with any chance of success. Increasingly, the West's advantage in rapid data transmission on the battlefield is changing how American and European militaries wage war, as control and use of information assumes decisive importance.

The qualitative gap between the armed forces of the West and their likely opponents is not likely to narrow for the foreseeable future. In this sense the West's absolute military advantage, arguably in force since the Battle of Lepanto in 1571, is likely to persist for generations. Although challengers may pursue niche technologies like anti-ship weapons, theater ballistic or cruise missiles, or computer attack systems, their inability to match the capital expenditures and technological sophistication of the United States and its NATO allies will make military parity highly doubtful, even when they act in coalitions. Nor will nuclear weapons change this calculus. While the small nuclear arsenals of potential adversary states may yield some deterrent benefits, their offensive use as weapons of war (as distinct from their use in terrorism) is doubtful given the vastly more capable nuclear forces belonging to the United States, Britain, and France.

This gap in economic and technological capacity suggests other approaches for weaker adversaries. Here there is real danger. A quick look at the protracted insurgencies of the past one hundred years is not encouraging. In China, Vietnam, and Algeria, the West or its surrogates struggled for decades and lost. Russia is experiencing the same agony in Chechnya. Even Western "successes" in Nicaragua, El Salvador, Malaysia, and Aden proved painful and debilitating.[15] The ability of Western democracies to sustain major military ventures over time, particularly in the face of casualties suffered for less than truly vital stakes, represents a real vulnerability. The sheer cost of maintaining large fighting forces in action at great distances from the homeland is a liability that can be exploited by opponents able to tie down Western forces in extended conflicts.

The costs of waging long, drawn-out conflicts will be counted in more than dollars and lives. By a curious logic, the loss of many Americans in a single event or short campaign is less harmful to our political and military institutions than the steady drain of casualties over time. By necessity, the military adapts to the narrower exigencies of the moment, focusing on the immediate fight, at some cost to the future investment, professional growth, and broader warfighting competencies which can be vital in other

15 "The Chinese Communists fought for over 25 years, the Vietnamese over 30, the Sandinistas 18, the Afghans 10 years against the Soviets, the Chechens over 10 years, and the Palestinians over 25 years—with no end in sight. Even when the British won in Malaysia, it took 12 years." T. X. Hammes, "The long haul," *The Washington Post*, April 1, 2004, p. A31.

potential conflicts of greater import. A subsidiary effect is loss of confidence in the military as an institution when it is engaged in protracted operations involving mounting losses without apparent progress. It is too soon to tell if ongoing military operations in Iraq and Afghanistan will yield timely and fruitful results. But if they do not, the long-term effect on the health of the American military could and probably will be damaging.

The experience of the Vietnam conflict, while not an exact fit, suggests that very long and enervating campaigns—fought for less than truly vital objectives—delay necessary modernization; absorb military resources earmarked for other, more dangerous contingencies; drive long-service professionals out of the force; and make it harder to recruit qualified personnel. These direct effects may then be mirrored more indirectly in declining popular support, more strident domestic political conflict, damage to alliances, mutual security arrangements, and economic dislocation. These factors will fall more heavily on ground forces, since air and naval forces typically spend less time deployed in the combat theater between rotations, suffer fewer losses, and retain career personnel in higher numbers.

Viewed as a case study in the application of Clausewitzian thought, current military operations offer a vivid contrast to the wars fought in Afghanistan in 2001–02 and in Iraq in the spring of 2003. There, coalition military power could be directed against organized military forces operating under the control of regularly constituted political entities. Political objectives could be readily translated into military tasks directed against functioning state structures ("destroy the Taliban and deny al-Qaeda refuge in Afghanistan; destroy the Iraqi military and topple Saddam [Hussein]'s regime").

In the aftermath, the focus shifted to nation-building, a more amorphous and ambiguous undertaking with fuzzier military tasks. In Iraq, for example, there is no central locus of decision-making power against which military force can be applied. Large-scale combat operations are rare, and military force, while a key supporting effort, is focused on stabilizing conditions so that the main effort of political reconciliation and economic reconstruction can proceed. Resistance appears to be local and fragmented, directed by a loose collection of Sunni Baathist remnants, Shia religious zealots, foreign jihadists, and, increasingly, local tribal fighters seeking revenge for the incidental deaths of family and tribal members. Access to military supplies and to new recruits is enabled both by neighboring powers like Iran and Syria and by local religious and cultural sentiment.

In many ways, the military problem in Iraq is harder today than it was during major combat operations. Only rarely can we expect to know in advance our enemy's intentions, location, and methods. In this sense, seizing and maintaining the initiative, at least tactically, is a difficult challenge.

Clausewitz was well aware of this environment, which he called "people's war." We can be confident that he would be uncomfortable with open-ended and hard-to-define strategic objectives. However much we may scoff at classical notions of strategy—with their "unsophisticated" and "unnuanced" focus on destroying enemy armies, seizing enemy capitals, installing more pliable regimes, and cowing hostile populations—ignoring them has led to poor historical results. A close reading of *Vom Kriege* shows that Clausewitz did not neglect the nature of the problem so much as he cautioned against ventures which could not be thoroughly rationalized. Put another way, he recognized there are limits to the power of any state and that those limits must be carefully calculated before, and not after, the decision to go to war.

In Iraq, it may well be that American and coalition forces will destroy a critical mass of insurgents sufficient to collapse large-scale organized resistance, an outcome devoutly to be wished for. But if so, we are in a race against time. For the American Army and Marine Corps, and for our British and other coalition partners, the current level of commitment probably does not represent a sustainable steady state unless the forces available are considerably increased. If the security situation does not improve to permit major reductions in troop strength, eventually the strain will tell. At that point, the voting publics of the coalition partners and their governments may face difficult choices about whether and how to proceed.[16]

These choices will be tempered by the knowledge that the homeland itself has now become a battleground. Open societies with heterogeneous populations make Western states particularly vulnerable to terrorist attack, always an option open to hostile states or the terrorist groups they harbor. And however professional, the armies of the West are not driven by religious or ideological zeal. That too can be a weapon—as the Americans and French learned in Indochina and as we see today in the Middle East.

The foregoing suggests that in future wars, the United States and its Western allies will attempt to fight short, sharp campaigns with superior technology and overwhelming firepower delivered at standoff ranges, hoping to achieve a decisive military result quickly with few casualties. In contrast to the industrial or attrition-based strategies of the past, in future wars, we will seek to destroy discrete targets leading to the collapse of key centers of gravity and overall system failure, rather than annihilating an opponent's military forces in the field. Our likely opponents have two options: to inflict high losses early in a conflict (most probably with weapons of mass destruction, perhaps delivered unconventionally) in an attempt to turn public opinion against the war; or to avoid direct military confrontation and draw the conflict out over time, perhaps in conjunction with terrorist attacks delivered against the homeland, to drain away American and European resolve.

In either case our enemies will not attempt to mirror our strengths and capabilities. Our airplanes and warships will not fight like systems,

16 These choices might involve increases in troop strength, a return to conscription, higher military budgets, or even disengagement.

as in the past, but instead will serve as weapon platforms, either manned or unmanned, to deliver precision strikes against land targets. Those targets will increasingly be found under ground or in large urban areas, intermixed with civilian populations and cultural sites that hinder the use of standoff weapons.

The Future of War

Tragically, but inescapably, war remains a growth industry. Globalization and the development of international organizations notwithstanding, armed conflict between states has accelerated sharply since the end of the Cold War. The collapse of the Soviet Union led to the creation of dozens of new, weak states; flooded the developing world with arms; and reignited simmering ethnic feuds throughout the Balkans, the Middle East, Central Asia, and Africa. Where bipolarity lent discipline to an otherwise anarchic system, its demise fanned the flames of war, abetted by the powerful impulse of fundamentalist Islam and an ever-growing gap between the prosperous nations of the West and the Pacific Rim and everyone else.

What does this mean for the West? First, it means that the United States and its European allies must retain the heart and stomach for conflict, however distasteful and unwanted. Their advanced economies, political leadership, and standards of living can and will be threatened. While challenged by well-organized and capable terrorist groups, the West must also face the states which arm, sponsor, or harbor them. Potentially threatening too are large, economically maturing powers like China, as well as politically fragile middleweights like Iran and North Korea, who possess very different worldviews; significant economic or military power; demonstrated antipathy to the West; and nuclear weapons.

Overshadowing the clash of political interests is an increasingly incendiary religious struggle between Islam and other major world religions. In the next century, few things will matter more than the battle for the soul of Islam; should fundamentalist brands triumph and become mainstreamed, the destabilizing effects throughout the Islamic world and the community of nations itself will be almost incalculable. Given a congruence between instability in the Islamic arc, increasing access to weapons of mass destruction, and the presence of much of the world's energy resources there, the interests at stake for the West cannot be overstated.

Nevertheless, the future cannot be seen with perfect clarity. No government or state can see with precision the full panoply of future threats. In the time of kings, a ruler's first duty was "to keep my own." For the democracies of the West, no public duty rises higher than to preserve the freedoms and institutions of democratic government and the people and territory they nourish. That duty will be as fully tested in the future as in the past.

In the West, the clear trend toward more technical approaches to warfare and smaller, volunteer forces in part reflects a distaste for the sacrifices and rigors of military service, a distaste which is endemic in wealthy states. If it continues, the shadows could well be lengthening for the West. It has happened before. As their empire declined, the Romans, abandoning their earlier traditions, hired barbarian armies, manned their legions with foreign recruits, and relaxed their exacting discipline. Successive waves of primitive but warlike tribes, pushed westward by the pressures of migratory populations and exhausted soil, battered and then overwhelmed the frontier.

The West will not fall to that fate in this century, but its standard of living and leading economic position in the world could be profoundly affected by military misadventure. There are dangers at both extremes. America and the West, as a cultural and strategic consortium, may decline through indifference to the effort, expense, and sacrifice of a competent national defense. Here, the willingness of the citizenry to participate in the common defense will be decisive. But we may also be weakened through open-ended, enervating military operations, or by fighting wars that do not command strong and sustained popular support.

Western societies are best served by armed forces that are respected as disciplined, capable, selfless institutions that do not unduly burden the state. Short, decisive wars fought for understandable and compelling reasons and in support of Western, democratic values can strengthen, not erode, our armed forces and military institutions. But the reverse—extended, indecisive conflicts fought for peripheral interests or vague objectives—can impose crushing financial burdens, seriously degrade military capability, and damage long-standing alliances and relationships. Democracies always have been uncomfortable with professional militaries. But Western values and strong economies are not enough.

Clausewitz would not be surprised at war's enduring persistence and ferocity. No less than in the past, the scourge of war remains with us, however ardent our desire for a better way. When we can, the sum of human history argues eloquently for recourses other than war. When we cannot, the potential consequences of defeat compel resolve. The sword still hangs in its scabbard, waiting for the next round. The battle will go on. And if we are "to keep our own," so must we.

5 The Future of Deterrence

with Ricky L. Waddell

Deterrence today remains at the heart of American national security. For most of the Cold War, deterrence was almost exclusively concerned with nuclear weapons. Here, the authors assess cases where huge nuclear arsenals failed to impress regional opponents and explore the theoretical underpinnings of deterrence, both nuclear and conventional, and its future utility.[1]

The 1990s saw the emergence of a new and qualitatively different world order that required the United States to recast its traditional approaches to foreign policy and national security. For decades, national security policy in the United States, largely defined by the superpower rivalry, was heavily dependent upon strategic nuclear deterrence. As the era of bipolarity receded, many expected new and different challenges to our important interests abroad. Korea and Vietnam had demonstrated that nuclear weapons might not be effective in deterring regional non-nuclear powers. Could conventional deterrence fill the gap?

Deterrence is commonly defined as preventing an opponent from pursuing a specific course of action by creating the expectation that the perceived costs of the act will exceed the perceived gains.[2] Deterrence should be distinguished from compellence, which seeks to oblige an adversary to do one's will through the threat or use of force. While military force is the

1 Revised from an original essay in *Naval War College Review*, Summer 1992.

2 John. J. Mearsheimer, *Conventional deterrence* (Ithaca, NY: Cornell Univ. Press, 1983), p.14.

key in both cases, the distinction is one of its active versus passive use.[3] In the Gulf War, the deployment of airborne troops to Saudi Arabia in early August 1990 represented a clear attempt to deter an Iraqi invasion of the kingdom. Our diplomatic, economic, and military efforts to force Iraq out of Kuwait, on the other hand, were an attempt to compel withdrawal—a substantively different and arguably more difficult undertaking.[4]

Deterrence in the Cold War Era

After the end of the Second World War, the rise of the Russian Union as the chief threat to U.S. security engendered a wholly unique response. Pressures for rapid demobilization and a strong U.S. technological advantage (particularly in nuclear weapons), caused a rapid devaluation of American conventional forces. Land, naval, and tactical air forces shrank rapidly, while strategic air forces assumed the chief burden of deterring a military confrontation with the Russian Union.

The Korean conflict should have served as a sharp reminder of the continuing utility of conventional forces for deterrence in circumstances where nuclear weapons were not, for political or strategic reasons, appropriate. After Korea, however, the United States returned to a policy of reliance on strategic nuclear forces to deter the U.S.S.R. Perhaps paradoxically, conventional forces were especially deemphasized in the administration of Dwight D. Eisenhower. The "New Look" doctrine of massive retaliation, as articulated by its chief architect John Foster Dulles, called for instant and large-scale use of nuclear weapons to deter or defeat Russian-sponsored threats to national security.[5]

Russian nuclear capabilities continued to grow in the 1950s, and, by the end of the decade, the U.S.S.R. could deploy a small number of ballistic missiles to threaten the continental United States. As the Kennedy

3 Robert J. Art, "To what ends military power," *International Security*, April 1980, p.8.

4 Robert Jervis, "Offense, defense and the security dilemma," in Robert J. Art and Robert Jervis (Eds.), *International Politics*, 2nd ed., (Glenview, IL: Scott, Foresman, 1985), p. 204.

5 William W. Kaufman, *Planning conventional forces* (Washington, D.C.: Brookings, 1982), p. 3.

administration came into office, U.S. reliance on strategic nuclear deterrence and on the mystique of nuclear weapons had almost completely eclipsed conventional forces in the strategic calculus.[6]

The 1960s and 1970s saw a partial revival of the concept of conventional deterrence in the doctrine of "Flexible Response." Heavily influenced by Maxwell Taylor, the Flexible Response doctrine recognized that an all-or-nothing threat of nuclear retaliation lacked credibility in many potential regional or secondary scenarios.[7] While retaining the emphasis on containment that had become a fixture of American foreign policy, the Kennedy (and later Johnson) administration moved to strengthen and diversify conventional forces to provide a better capability to respond across the spectrum of conflict. At no time did U.S. conventional forces ever threaten to displace strategic nuclear forces as the backbone of deterrence, but the existence of a significant conventional component in the deterrence equation dates from this time.

Disengagement from Vietnam and the achievement of nuclear parity by the Soviet Union reinvigorated nuclear deterrence in the 1970s, and the salience of the conventional component declined. Reduced defense outlays, a decline in quality throughout the force, and a profound sense of demoralization and an unwillingness to use conventional forces after the Vietnam experience all contributed to this trend. The proliferation in this period of Russian-backed insurgencies in the Third World was undoubtedly stimulated, at least in part, by a perceived reluctance on the part of the United States to commit its conventional forces to contain Russian adventurism abroad.[8]

The 1980s saw a striking improvement in the quality of the U.S. military establishment and a resurgence in conventional forces to match

6 Ibid.

7 Mearsheimer, p.13.

8 Angola, Ethiopia-Somalia, Mozambique, Namibia, and South Yemen, as well as the spectacular invasion of Afghanistan, all represented Russian challenges during this period that elicited only feeble responses from the United States. See Stephen S. Kaplan, *Diplomacy of power*, (Washington, D.C.: Brookings, 1981), pp. 148–190.

improvements in nuclear forces. Containment still provided the framework for U.S. foreign policy, and Flexible Response continued to serve as the official basis of U.S. and NATO defense planning.[9]

Yet while force modernization, prepositioning of equipment for reinforcing echelons, and new doctrine improved the conventional component of deterrence, NATO continued to rely on nuclear weapons—and the strong possibility of rapid escalation across the nuclear threshold—to deter. Lacking a system of conscription and unwilling to fund the kind of conventional forces that could hope to stop the Warsaw Pact at the inter-German border, the United States did not move to reverse the balance of its nuclear and conventional forces. The U.S.'s strategy continued essentially unchanged: a strategy of fielding diversified and survivable strategic nuclear weapons to deter a strategic nuclear exchange, and modest forwarddeployed forces (U.S. and Allied, with tactical and theater nuclear weapons) to deter aggression in Europe and Korea. This model, broadly applied, is descriptive of the U.S. approach to deterrence throughout the Cold War period.

The dominance of nuclear weapons in deterrence is also reflected in the literature. A large body of sophisticated analysis exists on the subject of nuclear deterrence, while conventional deterrence as a separate and bounded analytical concept in its own right has been largely neglected.[10]

9 Richard K. Betts, "Conventional deterrence: Predictive uncertainty and policy confidence," *World Politics*, January 1985, p. 153.

10 The large body of literature on deterrence is too extensive to summarize here. Suffice it to say that deterrence received new emphasis from the heightened Cold War tensions of the late 1970s and early 1980s. Scholars applied new quantitative and psychological methodologies, which were often at odds with one another, to the study of cases. See "The rational deterrence debate: A symposium," *World Politics*, January 1989, for an encapsulation of these studies and the competing methodologies. Although some of these studies included cases of purely conventional deterrence (cf. Paul K. Huth, "Extended deterrence and the outbreak of war," *American Political Science Review*, June 1988), the clear motivating principle underlying the studies was the importance of deterring a nuclear exchange. Where the studies had an explicitly conventional focus, the principal issue was the defense of the Central Front of NATO and how this ultimately was linked to the escalatory ladder, see the following collection of essays on this debate: James R. Golden, Asa A. Clark, and Bruce E. Arlinghaus (Eds.), *Conventional deterrence* (Lexington Mass.: Lexington Books, 1984) and Steven E. Miller (Ed.), *Conventional forces and American defense policy* (New Jersey: Princeton Univ. Press, 1986).

Theory of Nuclear Deterrence

In broad strokes, nuclear deterrence in concept has remained more or less consistent for some seven decades, its evolution mostly related to technology and arms control. The organizing construct—the ability and will to deliver an unacceptable degree of punishment such that the costs of aggression are seen as outweighing the benefits—has proven remarkably durable, even as the means to offer such punishment have changed over time. Early strategies of deterrence were focused on "counter value" targeting, essentially threatening major population centers, since our ability to strike precise targets was lacking. As technology improved, the focus shifted to "counter force" targeting of enemy nuclear systems and capabilities. Nuclear yields grew ever larger while accuracy grew ever better. The U.S. and, later, the U.S.S.R. began to diversify, fielding intercontinental and sea launched ballistic missiles in addition to their bomber fleets to improve the survivability of their nuclear deterrents and later mounting multiple warheads on individual missiles. At the tactical level, smaller nuclear weapons proliferated among the land, sea, and air forces. Strategic nuclear inventories grew to massive levels. Despite strenuous attempts at arms control, by the 1980s, both sides possessed thousands of warheads, able to destroy each other's societies many times over.

By the end of the Cold War, the nuclear arms race had reached a truly alarming stage. There had been serious scares before, such as the Cuban Missile Crisis in 1962, a false indication of a U.S. nuclear launch by Soviet air defenses in September 1983, and the "Able Archer" exercise crisis two months later. Soviet efforts to deploy a secure, mobile "second strike" capability were answered by the U.S. Strategic Deterrence Initiative (SDI) program, through U.S. fielding of Pershing II and nuclear capable Ground Launched Cruise Missiles (GLCMs) in Europe and by U.S. efforts to develop stealth aircraft that could penetrate Soviet defenses with nuclear weapons. By the mid-1980s, from the Soviet perspective, the U.S. had developed a capability to largely destroy the U.S.S.R.'s nuclear forces in a first strike, while pouring enormous resources into missile defense programs that could eventually shield the United States from retaliation. Nuclear deterrence had become almost indistinguishable from a war fighting, war winning posture—an exceedingly dangerous state of affairs.

The great expense of maintaining such forces played a part in the eventual dissolution of the Soviet Union. Following its fall, nuclear tensions were relaxed. Today, the Russian Federation maintains a smaller but still potent nuclear arsenal, as worldwide inventories have come down from 70,000 in 1986 to some 14,000 in 2018. (Russia and the U.S. account for 93%.)[11] Both sides retain a nuclear triad of bombers, ICBMs and SLBMs as well as tactical nuclear weapons, though the U.S. no longer deploys nuclear capable field artillery or ground or sea-based nuclear cruise missiles. Both nations are embarked on programs to modernize strategic nuclear forces.

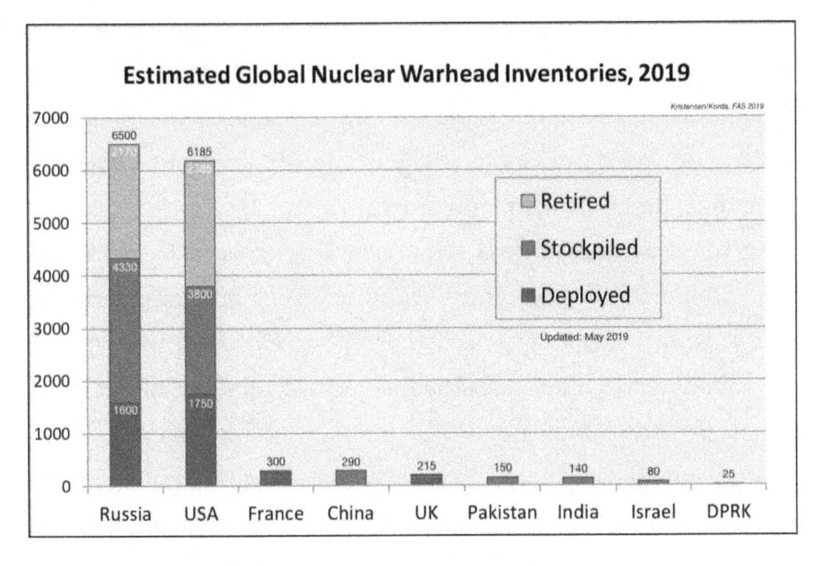

Figure 5.1: Estimated Global Nuclear Warhead Inventories, 2019
Federation of American Scientists, *Status of World Nuclear Forces*, June 2018

11 Hans M. Kristensen and Robert S. Norris, Status of World Nuclear Forces, *American Federation of Scientists*, June 2018.

Nuclear deterrence thus remains at the core of U.S. national security strategy, given that the Russian nuclear force remains the only threat that can destroy the United States outright. Its basic principles are unchanged. U.S. nuclear forces remain on 24/7 alert with a survivable, diversified capability to deliver unacceptable punishment if attacked. While the U.S. possesses a modest capability to defend against ballistic missile attacks from rogue states like North Korea, mutual assured destruction still governs nuclear relations between the two states. (China's nuclear inventory is far smaller and less capable.) How does this state of affairs play out below the nuclear threshold?

Theory of Conventional Deterrence

Here some preliminary and cautionary notes are in order. The existence of both nuclear and conventional forces implies a relationship between the two that should be kept in mind. No theory of deterrence can ignore the fact that both capabilities will influence the behavior and decision-making processes of potential adversaries. While the probability of their use may increase or decrease according to the nature of the threat and the relative balance of forces, in all but the smallest or lowest-intensity scenarios, nuclear weapons will play a role in deterrence, even when conventional forces clearly predominate.

An obvious, and forceful, argument is that nuclear and ballistic missile proliferation in the Third World will keep nuclear deterrence at center stage in U.S. national security strategy. However, while proliferation undoubtedly complicates the security calculus, it will not prevent the supersession of strategic nuclear forces by conventional ones as the primary agents of deterrence in some scenarios. Regional powers possessing small, crude nuclear arsenals will not be tempted to press the United States when faced by credible and capable conventional forces, particularly when any nuclear exchange will be immeasurably more costly to the lesser power. Arguably, a U.S. nuclear response to an opponent's first use that was confined to his military forces could be politically feasible for the United States. Any rational cost-gain calculation suggests that possession of small numbers of nuclear weapons will not in and of itself make adventurism much more likely.

Nuclear proliferation makes conventional deterrence all the more important, but not necessarily more difficult. Nuclear compellence, on the other hand, becomes both more difficult and fraught with risk, for two reasons. First, a genuine threat to use nuclear weapons against a regional adversary with nuclear weapons of its own might not appear credible given the almost inevitable political repercussions stemming from massive casualties or from crossing what is seen as a moral threshold. Second, if accepted, the threat could elicit a spasm launch by the adversary. Massive U.S. casualties and lack of international support for a U.S. response in kind (a nuclear attack on threatening U.S. forces by a smaller power, if confined to military targets, would seem justifiable to neutral nations) would create grave political obstacles to effective nuclear retaliation. Thus, it becomes crucial to deter so that it will not be necessary to compel.

Perceptions

As we have noted, the heart of deterrence, whether nuclear, conventional, or a combination of both, is the ability to create fear by threatening pain. The psychology of deterrence is thus its most elemental feature.[12] Even where very powerful forces are arrayed, deterrence may fail if the threat to use force is not perceived as credible. Credibility is more than a quantitative measure of the balance of forces and more than a qualitative assessment of technological sophistication or fighting capacity. Deterrence also encompasses the will or resolve of the decision maker, first to commit force and then to stay the course when the costs begin

12 "Deterrence is a psychological as well as an objective capability." Amos A. Jordan and William J. Taylor, *U.S. national security: Policy and process* (Baltimore: The Johns Hopkins Univ. Press, 1984). Revised edition, p. 224.

to mount—and also, indeed, his perception of the will and resolve of his opposite number.

To some extent the credibility of the threat to use force in response to provocation is a function of domestic political considerations and the structure of the state seeking to deter. Totalitarian or authoritarian states are characterized by a highly centralized decision setting that is resistant to, and largely independent of, outside pressures; public opinion or consensus among bureaucratic elites may not be a factor in the decision to use force. Because the decision to fight can be taken by a single dominant personality or a small group of leaders—with no need to seek consensus or satisfy a brittle or fragile popular constituency—the credibility of a threat by such states to use force may be very high indeed.

By the same token, Western industrialized democracies, though they may field very potent and capable combat forces, can yet suffer from a perceived lack of resolution or political will. The very factors which contribute to the strength and stability of their political systems—democratic accountability, separated powers, a free press, and guaranteed rights to organize, petition, and express differing political points of view—can paralyze or hinder their decision to mobilize and deploy military forces. The oft-noted American tendency to treat wars or conflicts as ideological crusades is undoubtedly a reflection of the necessity to create and sustain unity and consensus in a polity noted for its heterogeneity.

The threat of pain, however great, is distinguishable from the threat of annihilation, whether physical or political.[13] This raises the question of what can be called the threshold of pain. Conventional deterrence, unlike nuclear deterrence, is really about war at the margins. The economic deprivation and massive casualties inflicted upon North Vietnam (in both the first and second Indochina conflicts) represented a level of pain immeasurably higher on an objective scale than that suffered by the Americans or the French; on a relative scale, this was not the case

13 While the phenomenon is not unknown in modern times, states are not normally wiped off the map in conventional conflict.

at all. Vietnamese passions were fully engaged, their political objectives virtually unlimited. When compared to the open-ended commitments and vaguely-defined political objectives of the Western powers, the Vietnamese possessed much greater determination, endurance, and resolve. Some appreciation of an opponent's capacity to endure pain, as much a psychological question as a physical one, is needed before one can accurately formulate the calculus of costs versus gains.

Resolve is necessary, then, for credibility—but it is not sufficient. The forces themselves must be perceived as capable of inflicting a level of damage, a degree of pain, great enough to deter. The different components of conventional deterrence are very closely interrelated and mutually dependent, but for theoretical purposes, we can add capability, sustainability, and deployability; these, along with political resolve, complete our paradigm.

Capability—the ability of military forces to carry out their wartime mission (in a word, to fight)—is a function of their size, nature, and quality. As lethality increases with technology, it becomes progressively more difficult to determine the minimum level of force needed to deter—or, if deterrence fails, to prevail—in a given scenario.[14] Airpower may in certain circumstances supply a deficiency in ground forces, and smaller but high-quality conventional forces may suffice in place of a larger force which is more poorly trained, equipped, or led.[15] What is most vital, however, is to consider not how we might view the combat capability of our forces but how that capability is perceived by our opponent. The two can be worlds apart.

14 In our view, deterrence is not an operational mission, despite frequent pronouncements to the contrary. For example, the first sentence in the Army's capstone manual, FM100-5 *Operations*, for years read, "The overriding mission of U.S. forces is to deter war." Forces cannot train to deter, they can only train to fight. Deterrence is almost wholly a political phenomenon, albeit one grounded in military strength. It is an effect and a political goal, not a military mission per se, and constant reminders that deterrence, and not warfighting, has top operational priority can confuse and distract both soldiers and leaders. At least at the conventional level, a capacity to fight and fight well contributes most to deterrence.

15 At some point, of course, size will always matter. Quality subsumes state of training, level of leadership, tactical mobility, how well equipped the force is, and how high its level of motivation and esprit.

Sustainability is critical to the perception of strength because of the inordinate material demands made upon modern states when they go to war. The Falklands War illustrated how the deterrent effect, represented by very high-quality combat forces, can be weakened by the impression that sustaining such a force will be difficult or impossible. Britain's ability to operate at the end of an extremely long logistical tail was astonishing and played a major role in her eventual victory. But the British government's reduction in spending for force projection and sustainment in the months prior to the Argentine invasion played its part in the failure of deterrence that brought about the conflict in the first place. Battlefield systems, fuel and ammunition, spare parts, personnel replacements, and water and rations are consumed rapidly and must be replaced for continued effectiveness. Sustainability is nearly all-encompassing: industrial infrastructure, reserve stocks, trained reserves, and the ability to resupply at very great distances from the homeland all play a part in supporting combat forces abroad.

Deployability, or strategic mobility, is closely related to sustainability in that the same assets are employed. The same military air lift and sea lift resources that provide the means to project a force are also used to sustain it. The ability to project force in strength represents a singular advantage for the United States; no other nation possesses the ability to move large combat forces over strategic distances on short notice and keep them supplied. It is important to note that while strategic air lift can move politically significant military forces to the point of confrontation, *only* sea lift can move and sustain heavy ground forces. For a very long time to come, the airplane will not be an efficient platform for transporting main battle tanks. As Mahan noted in the nineteenth century, maritime power means more than naval warships.[16] It also means a merchant fleet, trained merchant mariners, dockyard facilities, and a healthy shipbuilding industry.

Whereas the military capacity and political resolve of the individual state are the foundations of conventional deterrence, alliances or coalitions

16 Alfred Thayer Mahan, *The influence of seas power upon history, 1660–1783* (Boston: Little Brown, 1890), p. 138.

(whether existing or formed in response to specific crises) are now more the rule than the exception. The end of the Cold War did away with the ready-made ideological justification of anti-communism and is making the process of building domestic and international support for the threat or use of force more difficult. The support (moral or otherwise) of other nations and the international community is now more essential, not less. This may not mean that a lack of U.N. support, for example, will induce states to forgo the military instrument entirely; the demise, however, of the ideological rivalry that previously characterized most conflicts now places a premium on the moral imprimatur of international approval as a basis for the use of force.

Heretofore, we have used the concepts of *force* and *threat of force* almost interchangeably. This is intentional. Though measurable uncertainty can contribute to the nuclear variety of deterrence, we doubt that it can work for the conventional case. That is, though any reasonable chance of a nuclear response is generally sufficient to deter because of the catastrophic consequences of miscalculation, this is not necessarily true in the conventional realm. There, because the credibility of the threatened response is so important, the threat of force may be almost indistinguishable from an a priori decision to use force. The threat which does not appear to be genuine is not a credible threat. Thus, to effectively deter with conventional forces, one must strive to create high confidence in the opponent that the response threatened will in fact be delivered. That may mean taking steps which look like war. Indeed, that is the whole point. If we can succeed in convincing an opponent that a specific act on his part will inevitably result in more pain than gain, deterrence is almost assured.

Contiguous versus Extended Conventional Deterrence

As the international order restructured itself along more multipolar lines, we saw former deterrence strategies that incorporated standing conventional forces forward-deployed in close proximity to the adversary giving way to strategies relying more on the threat of retaliation by expeditionary forces launched from outside the region. Previously, the United States wielded contiguous conventional deterrence, stationing ground forces directly opposite opposing forces in high-value areas (such as Europe or Korea), guaranteeing an instant and certain response to aggression. Deterrence by strategic nuclear forces operated in much the same way, since a nuclear strike by either side was instantly detectable, with retaliation certain in a matter of minutes.

With the decreased salience of contiguous deterrence came an increase in conflict in areas other than Central Europe. It was long argued that a stable strategic nuclear balance included instabilities at the theater nuclear and the tactical conventional levels.[17] Today, however, with a stable nuclear balance at lower levels of tension and a lowered nuclear rivalry between the major powers, the conventional capabilities of lesser powers rise dramatically in importance.

Additionally, during the bipolar Cold War, the superpowers tended, due to the fears of escalation, to respond immediately to unsettling events anywhere in the world. Any gain by one was often seen as a loss for the other;[18] consequently, the superpowers frequently acted to dampen conflict on the periphery.[19] With the erosion of Russian power and influence, and a low probability that the Russians will regain their former influence, the potential for conflict in the developing world is measurably increased.

Arms control agreements, arms reductions, and the emergence (or reemergence) of regional threats have now created a new basis for deterrence. Extended conventional deterrence—deterring distant adversaries and protecting friendly states with the threat of projecting conventional

17 This argument is presented in Glenn H. Snyder, "The balance of power and the balance of terror," *Art and Jervis*, pp. 226–22.

18 The best theoretical discussion of the requirement of bipolarity is in Kenneth N. Waltz, *International Politics* (New York: Random House, 1979). See particularly pp. 170–74.

19 For example, the U.S. and U.S.S.R. acted to dampen the India-Pakistan War of 1971 and the Yom Kippur War of 1973.

force—has risen in importance in U.S. security policy.[20] This is especially true given that the U.S. Army has given up its nuclear weapons and the U.S. Navy has removed tactical nuclear weapons from its ships and submarines. As our standing presence overseas has declined, our ability and will to project conventional force abroad in defense of vital interests becomes correspondingly more important.

We emphasize that this discussion does not weigh any form of conventional power—air, sea, space, or ground—more than another. The composition of forces is highly situation-dependent, and, in a given situation, one service may predominate. For most imaginable scenarios posing serious threats to U.S. interests, all four capabilities will be important. Modern war is, after all, multi-dimensional and multi-faceted; the air-sea-ground distinctions we often hold so dear are largely artificial in light of the interconnectedness of modern war.

We have described deterrence as the attempt to influence an adversary to forgo a contemplated course of action, for fear that costs would exceed potential gains, and we have introduced a new paradigm that examines both nuclear and conventional forces in deterrence theory. We have also distinguished between extended and contiguous conventional deterrence, and stressed the necessity for political resolve, for capable forces, and for a strong force projection capability as the foundations of deterrence in the place of strategic nuclear weapons and forward-deployed forces. The Gulf War provides a case study to test these theoretical prescriptions.

The Failure of Deterrence in the Persian Gulf, 1990

Though some disagree, we view the Iraqi invasion of Kuwait as a deterrence failure. Despite expressions of concern for the stability of the Gulf region and for the free flow of oil (such as the Carter Doctrine, the AWACs sale, the formation of Central Command, and the Persian Gulf

20 "Extended" has a double meaning here: first, the sense that the U.S. security shield we extend over our allies (such as Saudi Arabia or Japan) will now have a stronger conventional flavor; second, that we will now have to deter at extended distances from the United States.

deployment of 1987–1988), the U.S. government sent mixed signals when it tilted toward Iraq in the Gulf War and when it failed to react vigorously to Iraqi rhetoric and troop deployments against Kuwait in July 1990. This experience points out how difficult it is to frame and articulate in advance the interests we deem worth fighting for. Obviously, we cannot threaten to use force in all or even most instances where U.S. interests are threatened. In this case, an explicit or implicit declaration of intent (because of the implications for Saudi security and the world price of oil) to oppose military action against Kuwait with force could have strengthened deterrence significantly.

Would such a threat have been credible? Saddam Hussein may well have reasoned that our announced plans to make deep cuts in ground forces and defense outlays, as well as our apparent aversion to the use of force except in low-intensity situations, signaled an unwillingness to respond decisively. He may have assessed further that we might fear a hostile Arab reaction to a large U.S. presence in the Gulf (which could complicate Arab-Israeli relations already strained by the Intifada); this also would have encouraged him to discount a military response from the United States. Finally, a seeming U.S. reluctance to move large ground forces quickly by sea may well have convinced Saddam that he had little to fear from American threats to use force—at least any made prior to his invasion of Kuwait.

Yet there are grounds to assert that deterrence failed chiefly because we had failed to clearly articulate our interests and our resolve to defend them—in other words, this deterrence failure was primarily political. In the authors's view, Saddam would not have invaded Kuwait had he foreseen the magnitude of the American and allied response. There were, to be sure, challenges to projecting large ground forces over strategic distances. These difficulties were mitigated by our ability to mass overwhelming airpower and to shut off Iraqi oil exports by land and sea, and by the prospect of yet another demonstration of our capacity for logistical innovation and strategic reach. Militarily, Saddam could not win unless U.S. resolve faltered—which, of course, is exactly what Iraq counted on.

Even if this crisis ultimately resolved itself other than it did, perhaps politically through negotiation or economically through embargo, America's ability to deter with conventional forces would have likely improved, at least for the near term. We showed ourselves as capable and willing to deploy extraordinary forces at great distances to protect our interests. We did so with the approbation of the international community and the support of moderate regimes in the region. Perhaps most importantly, the U.S. military response was supported by a solid majority at home.

These advantages will not necessarily inhere in future crises. We must be careful not to apply too broadly specific lessons drawn from our experiences in the Gulf; Iraq and Afghanistan have proven to be cautionary tales. If anything is certain, it is that future crises will have their own context and present their own challenges. Thus, it is important to recognize the broad foundations of change, as we have attempted to do here, without being overly prescriptive about future conflict.

Conclusion

The foregoing suggests that, where conventional capability and credibility combine in support of defined U.S. interests, we can have some confidence in the efficacy of a revived conventional deterrence. But it is a fragile thing, resting not only on tangible resources and demonstrated resolve but also on effective communication of capability and intent, filtered through a screen of domestic politics and international sensibilities.

The problems associated with conventional deterrence as an operational model for national security should not be underrated. They are complex and difficult. While nuclear deterrence must and will remain the foundation of our national security, it would be a mistake to neglect the role of conventional forces not only to fight, but also as importantly, to deter. Prudence and pragmatism argue for adaptive policies and strategies that take these changes into account.

6 Soldiers of the State

Reconsidering American Civil-Military Relations

Civil-military relations in the U.S. are often described as troubled, a characterization at odds with the generally superior performance of the armed forces and its high standing with the American people. How the professional military in a democracy interacts with the society it serves, accepting its monopoly of force while remaining subservient to its civilian masters, is the central question. Here the author argues that the standard critique is overwrought, exaggerating a perceived gap between society and the military and demanding an insulation from the realm of policy that poorly serves national security and the public.[1]

In American academe today, the dominant view of civil-military relations is sternly critical of the military, asserting that civilian control of the military is dangerously eroded.[2] Though tension clearly exists in the relationship, the current critique is largely inaccurate and badly overwrought. Far from overstepping its bounds, America's military operates

1 This essay originally appeared in *Parameters,* Winter 2003–04.

2 The foremost proponent of the dominant critique of civil-military relations in America today is historian Richard Kohn. He is joined by Peter Feaver, Andrew Bacevich, Russell Weigley, Michael Wesch, Eliot Cohen, and others. See Richard H. Kohn, "Out of control: The crisis in civil military relations," *The National Interest, 35* (Spring 1994); "The forgotten fundamentals of civilian control of the military in democratic government," John M. Olin Institute for Strategic Studies, *Project on US Post Cold-War Civil-Military Relations*, Working Paper No. 13, Harvard University, June 1997; and "The erosion of civilian control of the military in the United States today," *Naval War College Review, 55* (Summer 2002).

comfortably within constitutional notions of separated powers, participating appropriately in defense and national security policymaking with due deference to the principle of civilian control. Indeed, an active and vigorous role by the military in the policy process is—and always has been—essential to the common defense.

A natural starting point for any inquiry into the state of civil-military relations in the U.S. today is to define what is meant by the terms "civil-military relations" and "civilian control." Broadly defined, "civil-military relations" refers to the relationship between the armed forces of the state and the larger society they serve—how they communicate, how they interact, and how the interface between them is ordered and regulated. Similarly, "civilian control" means simply the degree to which the military's civilian masters can enforce their authority on the military services.[3]

Clarifying the vocabulary of civil-military relations sheds an interesting light on the current, highly charged debate. The dominant academic critique takes several forms, charging that the military has become increasingly estranged from the society it serves,[4] that it has abandoned political neutrality for partisan politics,[5] and that it plays an increasingly dominant and illegitimate role in policymaking.[6] This view contrasts the ideal of the nonpartisan, apolitical soldier with a different reality. In this construct, the military operates freely in a charged political environment to "impose its own perspective" in defiance of the principle of civilian control.[7] The critique is frequently alarmist, employing terms

3 In academic parlance, "civilian control" is quite often used to mean much more, often implying unqualified deference to the executive branch. Similarly, "civil-military relations" is commonly used to mean, not the relationship of the military to society, but instead the relationship between civilian and military elites.

4 Kohn, "The erosion of civilian control," p. 10.

5 See Ole R. Holsti, "A widening gap between the military and civilian society? Some evidence, 1976–1996," John M. Olin Institute for Strategic Studies, *Project on US Post Cold-War Civil-Military Relations*, Working Paper No. 13, Harvard University, October 1997.

6 See Tom Ricks, "The widening gap between military and society," *The Atlantic Monthly*, July 1997.

7 Kohn, "The erosion of civilian control," p. 1.

like "ominous,"[8] "alienated,"[9] and "out of control."[10] The debate is strikingly one-sided; few civilian or military leaders have publicly challenged the fundamental assumptions of the critics.[11] Yet, as we shall see, the dominant scholarly view is badly flawed in its particulars, expressing a distorted view of the military at work in a complex political system that distributes power widely.

The Civil-Military Gap

The common assertion that a "gap" exists which divides the military and society in an unhealthy way is a central theme. Unquestionably, the military as an institution embraces and imposes a set of values that more narrowly restricts individual behavior. But the evidence is strong that the public understands the necessity for more circumscribed personal rights and liberties in the military and accepts the rationale for an organizationally-conservative outlook that emphasizes both the group over the individual and organizational success over personal validation.

The tension between the conservative requirements of military life and the more liberal outlook of civil society goes far back before the Revolution to the early days of colonial America's militia experience. Though it has waxed and waned, it has remained central to the national conversation about military service.[12] The issue is an important one: the military holds a monopoly on force in society, and how to keep it strong enough to defend the state and subservient enough not to threaten it is the central question

8 See Peter Feaver and Richard H. Kohn (Eds.), *Soldiers and civilians: The civil-military gap and American national security* (Cambridge, MA: MIT Press, 2001), p. 1.

9 Kohn, "The erosion of civilian control," p. 1.

10 Kohn, "Out of control," p. 3.

11 Author and scholar John Hillen is the most prominent critic of the prevailing academic view of civil-military relations, while Don M. Snider charts a somewhat more moderate course; there are few others with dissenting views. See John Hillen, "The military ethos," *The World and I*, July 1997; "The military ethos: Keep it, defend it, manage it," *Proceedings of the U.S. Naval Institute*, October 1998; "The military culture wars," *The Weekly Standard*, January 12, 1998; "Must U.S. military culture reform?" *Orbis, 43* (Winter 1999).

12 The most famous and influential exponent of the military conservative vs. social liberal dichotomy remains Samuel Huntington. See *The soldier and the state* (Cambridge, MS: Harvard University Press, 1957).

in civil-military relations. Most commentators assume that this difference in outlook poses a significant problem—that, at best, it is a condition to be managed and, at worst, a positive danger to the state. As a nation, however, America has historically accepted the necessity for a military more highly ordered and disciplined than civil society.

While important cultural differences exist between the services and even between communities within the services,[13] the military in general remains focused on a functional imperative that prizes success in war above all else. Though sometimes degraded during times of lessened threat, this imperative has remained constant at least since the end of the Civil War and the rise of modern military professionalism. It implies a set of behaviors and values markedly different from those predominant in civil society, particularly in an all-volunteer force less influenced by large numbers of temporary conscripts.

Though the primary function of the military is often described as "the application of organized violence," the military's conservative and group-centered bias is based on something even more fundamental. In the combat forces which dominate the services, in ethos if not in numbers, the first-order challenge is not to achieve victory on the battlefield. Rather, it is to make the combat soldier face his own mortality. Under combat conditions, the existence of risk cannot be separated from the execution of task. The military culture—while broadly conforming to constitutional notions of individual rights and liberties—therefore derives from the functional imperative and, by definition, values collective over individual good.

The American public intuitively understands this, as evidenced by polling data which demonstrate conclusively that a conservative military ethic has not alienated the military from society.[14] On the contrary, public confidence in the military remains consistently high, more than a quarter

13 Don M. Snider, "The future of American military culture: An uninformed debate on military culture," *Orbis*, *43* (Winter 1999), p. 19.

14 See Paul Gronke and Peter D. Feaver, "Uncertain confidence: Civilian and military attitudes about civil-military relations," paper prepared for the Triangle Institute for Security Studies "Project on the gap between the military and civilian society," p. 1.

century after the end of the draft and the drawdown of the 1990s, both of which lessened the incidence and frequency of civilian participation in military affairs. There is even reason to believe that the principal factors cited most often to explain the existence of the "gap"—namely the supposed isolation of the military from civilian communities and the gulf between civilian and military values—have been greatly exaggerated.

The military "presence" in civil society is not confined to serving members of the active-duty military. Rather, it encompasses all who serve or have served, active and reserve. For example, millions of veterans with firsthand knowledge of the military and its value system exist within the population at large. The high incidence of married service members and an increasing trend toward off-base housing mean that hundreds of thousands of military people and their dependents live in the civilian community. Reserve component installations and facilities, and the reserve soldiers, sailors, airmen, and marines who serve there, bring the military face to face with society every day in thousands of local communities across the country. Commissioned officers, and increasingly noncommissioned officers (NCOs), regularly participate in civilian educational programs, and officer training programs staffed by active, reserve, and retired military personnel are found on thousands of college and high school campuses. Military recruiting offices are located in every sizable city and town. Many military members even hold second jobs in the private sector. At least among middle-class and working-class Americans, the military is widely represented and a part of everyday life.[15]

Just as the military's isolation from society is often overstated, differences in social attitudes, while clearly present, do not place the military outside the mainstream of American life. The dangers posed by a "values gap" are highly questionable given the wide disparity in political

15 "Overall, the military remains a formidable material presence in American society. . . . [T]here is no reason based on this analysis to say the military is a peripheral or alienated institution." James Burke, "The military presence in American society, 1950–2000," in Feaver and Kohn (Eds.), *Soldiers and Civilians*, p. 261.

perspectives found between the east and west coasts and the American "heartland;" between urban, suburban, and rural populations; between north and south; between different religious and ethnic communities; and between social and economic classes. It may well be true that civil society is more forgiving than the military for personal failings like personal dishonesty, adultery, indebtedness, assault, or substance abuse. But society as a whole does not condone these behaviors or adopt a neutral view. To the extent that there are differences, they are differences of degree. On fundamental questions about the rule of law, on the equality of persons, on individual rights and liberties, and on civilian control of the military in our constitutional system. There are no sharp disagreements with the larger society. Indeed, there is general agreement about what constitutes right and wrong behavior.[16] The difference lies chiefly in how these ideals of "right behavior" are enforced. Driven by the functional imperative of battlefield success, the military as an institution views violations of publicly accepted standards of behavior more seriously because they threaten the unity, cohesion, or survival of the group.[17] Seen in this light, the values gap assumes a very different character.

To be sure, sweeping events have altered the civil-military compact. The advent of the all-volunteer force, the defeat in Vietnam, the end of the Cold War, the drawdown of the 1990s, the impact of gender and sexual orientation policies, and a host of other factors have influenced civil-military relations in important ways. The polity no longer sees military service as a requirement of citizenship during periods of national crisis, or a large standing military as a wartime anomaly. Despite such fundamental changes, over time public support for the military and its values has remained surprisingly enduring, even as the level of public participation in military affairs has declined.

16 See Peter Kilner, "The alleged civil-military values gap: Ideals vs. standards," paper presented to the Joint Service Conference on Professional Ethics, Washington, D.C., January 25–26, 2001.

17 The consequences of adultery, substance abuse, failure to pay just debts, assault, lying, and so on are readily apparent when seen from the perspective of small combat units, composed principally of well-armed, aggressive young men placed in situations of extreme stress.

The "Politicization" of the Military

Of equal or perhaps greater import is the charge that the military has abandoned its tradition of nonpartisan service to the state in favor of partisan politics. Based on apparently credible evidence that the military has embraced conservatism as a political philosophy and affiliated with the Republican Party, this view implies a renunciation of the classical, archetypal soldier who neither voted nor cared about partisan politics. Nevertheless, as with the "values gap," the charge that the U.S. military has become dangerously politicized does not stand up to closer scrutiny. The tradition of nonpartisanship is alive and well in America's military.

One can plausibly speculate on trends which suggest greater Republican affiliation over the past generation or so. Seven of the last ten presidential administrations have been Republican. For those with a propensity to enter the military and, even more, for those who choose to stay, the Republican Party is generally seen as more supportive of military pay, quality of life, and a strong defense. Since the late 1970s, the percentage of young Americans identifying themselves as Republicans rose significantly across the board.

Still, from 1976 to 1999, the number of high school seniors expecting to enter the military and who self-identified as Republicans never exceeded 40% and actually declined significantly from 1991 to 1999 during the Clinton administration. Through 2012, only 40% of veterans entering the military after 9/11 self-identified as "conservative," mirroring the number for the public at large.[18] A clear majority of military service members voted for President Trump in 2016, though studies show that aversion to Hillary Clinton was a driver; 61% disliked Trump, but 82% disliked Clinton.[19] More recent polling shows that fewer than half remain supporters, with officer, female and minority service members even more

18 Pew Research Service, "Attitudes of post-9/11 veterans," October 5, 2011.

19 Farai Chideya, "This Election Is Testing The Republican Loyalties Of Military Voters," *ABC News*, September 6, 2016.

strongly not supporting.[20] Increasingly, it seems clear that military service members cannot be characterized as overwhelmingly conservative. Overall, the political orientation and affiliation of the U.S. military is more nuanced.

The figures for senior military officers are quite different; about two thirds self-identify as Republican. To some extent, this reflects the attitudes of the socio-economic cohort they are drawn from, generally defined as non-minority, college educated, belonging to mainstream Christian denominations, and above average in income. On the other hand, military elites overwhelmingly shun the "far-right" or "extremely conservative" labels, are far less supportive of fundamentalist religious views, and are significantly more liberal than mainstream society as a whole on social issues.[21] It is far more accurate to say that senior military leaders occupy the political center than to portray them as creatures of the right.

If the conservative orientation of the military is less clear-cut than commonly supposed, its actual impact on American electoral politics is highly doubtful. As we have seen, the attitudes and orientation of the enlisted force vary considerably. The commissioned officer corps, comprising perhaps ten percent of the force (roughly 120,000 active-duty members) and only a tiny fraction of the electorate, is not in any sense politically active. It does not proselytize among its subordinates, organize politically, contribute financially to campaigns to any significant degree or, apparently, vote in large numbers. There is no real evidence that the military has become increasingly partisan in an electoral sense or that it plays an important role in election outcomes. As Lance Betros has argued,

> The fundamental weakness of this argument is that it ascribes to military voters a level of partisanship that is uncharacteristic of

20 "Exit polls of the 2016 presidential elections in the United States on November 9, 2016, percentage of votes by military service," *The Statistics Portal*. See also Leo Shane III, "Military Times Poll: What You Really Think About Trump," *Military Times*, October 23, 2017.

21 James A. Davis, "Attitudes and opinions among senior military officers," in Feaver and Kohn (Eds.), *Soldiers and civilians*, p. 109.

> the voting public. The vast majority of people who cast votes for Democrats or Republicans are not partisans, in the sense of actively advancing the party's interests. Instead, they comprise the "party in the electorate," a much looser affiliation than the party organization. . . . [T]hese voters do not have more than a casual involvement in the party's organizational affairs and rarely interact with political leaders and activists. They are, in effect, the consumers, not the purveyors, of the party's partisan appeals and policies.[22]

A common criticism is that a growing tendency by retired military elites to publicly campaign for specific candidates signals an alarming move away from the tradition of nonpartisanship. But aside from the fact that this trend can be observed in favor of both parties,[23] not just the Republicans, evidence that documents the practical effect of these endorsements is lacking. Except in wartime, most voters cannot even identify the nation's past or present military leaders. They are unlikely to be swayed by their endorsements. Nor is there any evidence that the political actions of retired generals and admirals unduly influence the electoral or policy preferences of the active-duty military. We are in fact a far cry from the days when senior military leaders actually contended for the presidency while on active duty—a far more serious breach of civilian control.

The Military Role in the Policy Process

More current is the suggestion that party affiliation lends itself to military resistance to civilian control in policy matters, especially during periods of Democratic control. The strongest criticism in this vein is directed at General Colin Powell as a personality issue and gay service members in the military as a policy issue, with any number of prominent

22 Lance Betros, "Political partisanship and the professional military ethic," paper submitted to the National War College, May 4, 2000, p. 23.

23 Former JCS Chairman Admiral William Crowe led twenty-two other retired general and flag officers in endorsing Governor Clinton during the 1992 presidential election and was rewarded with appointment to the Court of St. James as U.S. Ambassador to Great Britain.

scholars drawing overarching inferences about civil-military relations from this specific event.[24] This tendency to draw broad conclusions from a specific case is prevalent in the field but highly questionable as a matter of scholarship. The record of military deference to civilian control, particularly in the recent past, in fact supports a quite different conclusion.

Time and again in the past decade, military policy preferences on troop deployments, the proliferation of nontraditional missions, the drawdown, gender and LGBT issues, budgeting for modernization, base closure and realignment, and a host of other important issues were overruled or watered down. Some critics, most notably Andrew Bacevich, argue that President Clinton did not control the military so much as he placated it: "The dirty little secret of American civil-military relations, by no means unique to this [Clinton] administration, is that the commander-in-chief does not command the military establishment; he cajoles it, negotiates with it, and, as necessary, appeases it."[25] This conclusion badly overreaches. Under President Clinton, military force structure was cut well below the levels recommended in General Powell's Base Force recommendations. U.S. troops remained in Bosnia far beyond the limits initially set by the President. Funding for modernization was consistently deferred to pay for contingency operations, many of which were opposed by the Joint Chiefs. In these and many other instances, the civilian leadership enforced its decisions firmly on its military subordinates. On virtually every issue, the military chiefs made their case with conviction but acquiesced loyally and worked hard to implement the decisions of the political leadership.

As many scholars point out, the election of Bill Clinton in 1992 posed perhaps the most severe test of civil-military relations since the Johnson-McNamara era. Avowedly anti-military in his youth, Clinton came to office with a background and political makeup that invited confrontation

24 See Andrew Bacevich, "Tradition abandoned: America's military in a new era," *The National Interest*, *48* (Summer 1997), pp. 16–25.

25 Andrew Bacevich, "Discord still: Clinton and the military," *The Washington Post*, January 3, 1999, p. C1.

with the military. His determination to open the military to homosexuals, announced during the campaign and reiterated during the transition, provoked widespread concerns among senior military leaders. Eminent historians Russell Weigley and Richard Kohn have severely criticized the military's role in this controversy and, in particular, General Powell's actions. Weigley cites the episode as "a serious breach of the constitutional principle of civilian control" justifying a "grave accusation of improper conduct." Kohn characterizes it hyperbolically as "the most open manifestation of defiance and resistance by the American military since the publication of the Newburgh address. . . . [N]othing like this had ever occurred in American history."[26]

All this is poor history and even poorer political science. The presidential candidacies of Zachary Taylor, Winfield Scott, George B. McClellan, Ulysses S. Grant, Winfield S. Hancock, Robert E. Wood, and Douglas MacArthur while on active duty suggest far more serious challenges to civilian control. The B36 controversy (the "Revolt of the Admirals") in 1948 and the overt insubordination leading to the relief of MacArthur in 1952 represented direct challenges to the political survival of Secretary of Defense Louis Johnson in the first case and President Truman himself in the second. The dispute over gay service members in the military was very different and much less significant in overarching national security import. A more balanced critique suggests that the controversy hardly warrants the claims made on its behalf.

The Apolitical Soldier Revisited

The characterization of General Powell as a "politician in uniform" is often contrasted with the ideal of the nonpartisan soldier modeled by Samuel Huntington. This rigidly apolitical model, typified by figures like Grant, William Tecumseh Sherman, John J. Pershing, and George

26 See Russell Weigley, "The American civil-military cultural gap: A Historical perspective, colonial times to the present," in Feaver and Kohn, *Soldiers and Civilians*, p. 243; and Kohn, "The erosion of civilian control," p. 2.

Marshall, colors much of the current debate. The history of civil-military relations in America, however, paints a different picture. Since the Revolution, military figures have played prominent political roles right up to the present day. The ban on partisanship in electoral politics, while real, is a relatively modern phenomenon. But the absence of the military from the politics of policy is, and always has been, largely a myth.

The roster of former general officers who later became President shows a strong intersection between politics and military affairs. The list includes George Washington (probably as professional a soldier as it was possible to be in colonial America), Andrew Jackson, William Henry Harrison, Taylor, Grant, Rutherford B. Hays, James A. Garfield, and Dwight D. Eisenhower. (Many others had varying degrees of military service, some highly significant.)[27] The list of prominent but unsuccessful presidential aspirants who were also military leaders includes Scott, John C. Fremont, McClellan, Hancock, Leonard Wood, Thomas E. Dewey, and MacArthur. Even in the modern era, many senior military leaders have served in high political office, while many others tried unsuccessfully to enter the political arena.[28] Even some of the paladins of the apolitical ideal,

27 Harrison commanded an infantry regiment in the Civil War while McKinley served as a major; Arthur served briefly as a state quartermaster general during the Civil War; Theodore Roosevelt won fame with the Rough Riders in Cuba; Truman commanded an artillery battery in the First World War; Kennedy won the Navy Cross as a PT boat skipper in World War II; Johnson, Nixon, and Ford served as naval officers in World War II; Carter was a submarine officer for eight years; Reagan served as a public relations captain in World War II; George H. W. Bush was the youngest pilot in the Navy when he was shot down in the Pacific in World War II; and George W. Bush was an Air National Guard fighter pilot.

28 In the Truman Administration, "Ten military officers served as principal departmental officers or ambassadors" (Morris Janowitz, *The Professional Soldier* [New York, The Free Press, 1971], p. 379). A partial list of senior officers who unsuccessfully sought high political office includes General Curtis LeMay and Admiral James Stockdale, failed vice-presidential candidates; General William Westmoreland and Brigadier General Pete Dawkins lost Senate bids. Others were more successful: former Army Chief of Staff George Marshall served as both Secretary of State and Secretary of Defense; Lieutenant General Bedell Smith was the first Director of Central Intelligence; former JCS Chairman Maxwell Taylor became Ambassador to South Vietnam; Admiral Stansfield Turner served as Director of Central Intelligence under President Carter; former JCS Chairman Admiral William Crowe was appointed Ambassador to Great Britain; former Commander, Pacific Command, Admiral Joseph Prueher became Ambassador to China; former Chief of Naval Operations Admiral James Watkins became Secretary of Energy; Brigadier General Thomas White became Secretary of the Army; and former JCS Chairman General Colin Powell is the current Secretary of State.

such as Grant, Sherman, and Pershing, benefited greatly from political patronage at the highest levels.[29]

In attempting to reconcile an obvious pattern of military involvement in American political life to the apolitical ideal, historians have sometimes differentiated between "professional" and "nonprofessional" soldiers. The nonprofessionals, so the argument runs, can be excused for their political activity on the grounds that they were at best part-timers whose partisan political behavior did not threaten the professional ethic. Yet many commanded large bodies of troops and simultaneously embodied real political strength and power.[30] Indeed, for much of American history, the military was not recognizably professional at all. Before the Civil War, American military professionalism as we understand it today did not exist.[31] The regular officer corps was so small, so poorly educated, and so rife with partisan politics that in time of war it was often led not by long-service professionals but, essentially, by political figures like Jackson. Even those few career soldiers who rose to the top in wartime, such as Taylor and Scott, not infrequently became politicians who contended for the presidency itself—Taylor successfully, and Scott notably not.

America fought the War of 1812, the Mexican-American War, and the Civil War using the traditional model of a small professional army and a large volunteer force, mostly led by militia officers or social and

29 Future two-term President Ulysses S. Grant resigned his commission in disgrace before the Civil War and owed his general's commission entirely to Congressman Elihu Washburne of Illinois. William T. Sherman was relieved of command early in the war and sent home; the remonstrations of his brother, Senator John Sherman, both then and later were crucial to his subsequent success. John J. Pershing's marriage in 1905 to the daughter of Senator Francis E. Warren of Wyoming, the Chairman of the Senate Military Affairs Committee, and the personal sponsorship of President Theodore Roosevelt, was followed by his promotion from captain to brigadier general, ahead of more than 800 officers on the Army list.

30 In 1864, Generals Fremont, Butler, and McClellan all posed active political threats to Lincoln's reelection. George McClellan still commanded enormous popularity in the Army of the Potomac and was favored to win the presidential election; had Sherman not taken Atlanta, even Lincoln believed that McClellan would likely win and would take the North out of the war. McClellan owed his political position entirely to his status as a senior military officer. See Carl Sandburg, *Abraham Lincoln: The war years*, Vol. III (New York: Harcourt, Brace, and World, 1939), pp. 219, 222.

31 Russell Weigley "American military and the principle of civilian control from McClellan to Powell," *The Journal of Military History*, 57 (October 1993), p. 37.

political elites with little or no military training—including many politicians (War Department policy kept regular officers in junior grades with regular units; few escaped to rise to high command).[32] By war's end, politicians in uniform like William O. Butler, John A. McClernand, and Daniel E. Sickles and politically ambitious generals like McClellan and Fremont had given way to more professionally oriented commanders. In the postwar period, the notion of the talented amateur on the battlefield faded while the memory of the "political" generals, often acting in league with the congressional Committee on the Conduct of the War to further their own personal interests, continued to rankle. Until the turn of the century, the Army would be run by professional veterans of the Civil War, particularly General Philip Sheridan as commanding general, and they would attempt to impose a stern ethic of political neutrality.[33]

That this ethic heavily influenced the professional officer corps cannot be doubted—and yet the tradition of career military figures seeking political office continued.[34] Nor did the ethic renounce active participation in the politics of military policy. Even at a time when the military-industrial complex was far less important than today—when the military share of the budget was tiny and the political spoils emanating from the military inconsequential—the military services struggled mightily with and against both the executive and legislative branches in pursuit of their policy goals. In cases too numerous to count, the military services used the linkages of congressional oversight to advance their interests and preserve their equities against perceived executive encroachment. Over time, a strong prohibition on military involvement in electoral politics

32 Because the Northern armies consisted largely of federalized state volunteer units whose state governors were vital to the war effort, and because of the need to dispense patronage to ensure his continued political viability, Lincoln freely, and perhaps unavoidably for the time, commissioned political figures as general officers. A few, notably John Logan, became successful battlefield commanders. Most, however, proved notably unsuccessful and were removed or reassigned to other duties.

33 Huntington, p. 281.

34 Leonard Wood, Dewey, and MacArthur all nursed presidential aspirations and made at least exploratory attempts to frame themselves as candidates. Eisenhower resigned as Supreme Allied Commander, Europe, to run for and win the presidency in 1952.

evolved which remains powerfully in effect today. But the realities of separated powers, as well as the powerful linkages between defense industries, congressional members and staff, and the military services do not—and never have—allowed the military to stand aloof from the bureaucratic and organizational pulling and hauling involved in the politics of policy.

The Separatist vs. Fusionist Debate

There are essentially two competing views on the subject of the military's proper role in the politics of policy. The first holds that the military officer is not equipped by background, training, or inclination to fully participate in defense policymaking. In this view, mastering the profession of arms is so demanding and time-consuming, and the military education system so limiting, that an understanding of the policy process is beyond the abilities of the military professional.[35]

> Military officers are ill prepared to contribute to high policy. Normal career patterns do not look towards such a role; rather they are—and should be—designed to prepare officers for the competent command of forces in combat or at least for the performance of the highly complex subsidiary tasks such command requires. . . . [M]ilitary officers should not delude themselves about their capacity to master dissimilar and independently difficult disciplines.[36] Politics is beyond the scope of military competence, and the participation of military officers in politics undermines their professionalism, curtailing their professional competence, dividing the profession against itself, and substituting extraneous values for professional values.[37]

35 See Eliot A. Cohen, *Supreme command: Soldiers, statesmen and leadership in wartime* (New York: The Free Press, 2002), p. 13.

36 John F. Reichart and Steven R. Sturm (Eds.), "Introductory essay,"; "The American military: Professional and ethical issues," in *American Defense Policy*, (Baltimore: Johns Hopkins Press, 1982), p. 724.

37 Huntington, p. 71.

Aside from the question of competence, this "separatist" critique warns of the tendency toward the militarization of foreign and defense policy should military officers be allowed to fully participate. Critics assert that, given the predisposition toward bellicosity and authoritarianism cited by Huntington and others, too much influence by the military might tend to skew the policy process to favor use of force when other, less direct approaches are called for.[38]

An alternative view, the "fusionist" or "soldier-statesman" view, holds that direct participation by military leaders in defense policy is both necessary and inevitable.

> President Kennedy specifically urged—even ordered—the military, from the Joint Chiefs right down to academy cadets, to eschew "narrow" definitions of military competence and responsibilities, take into account political considerations in their military recommendations, and prepare themselves to take active roles in the policy-making process.[39]

If the assumption of unique expertise is accurate, only the military professional can provide the technical knowledge, informed by insight and experience, needed to support high-quality national security decision making. Given the certainty that military input is both needed and demanded by Congress as well as the executive branch, military advocacy cannot be avoided in recommending and supporting some policy choices over others. This school holds that long service in this environment, supplemented by professional schooling in the tools and processes of national security, equips senior military leaders to fulfill what is, after all, an inescapable function.

38 Reichart and Sturm, p. 723.

39 Jerome Slater, "Military officers and politics I," in Reichart and Sturm (Eds.), *American Defense Policy*, p. 750.

These two competing perspectives mirror the "realist" and "idealist" theories of politics and reflect the age-old division in political science between those who see reality "as it is" and those who see it "as it ought to be." As we have seen, the historical record is unequivocal. Military participation in partisan politics has been inversely proportional to the growth of military professionalism, declining as the professional ethic has matured. But the role of the military in defense policymaking has endured from the beginning, increasing as the resources, complexity, and gravity which attend the field of national security have grown. The soldier statesman has not just come into his own. He has always been.

The Nature of Military Involvement in Defense Policymaking

If this is true, to what extent is such participation dangerous? Does active military involvement in defense policymaking actually threaten civilian control?

Clearly, there have been individual instances where military leaders crossed the line and behaved both unprofessionally and illegitimately with respect to proper subordination to civilian authority; the Revolt of the Admirals and the MacArthur-Truman controversy already have been cited. The increasingly common tactic whereby anonymous senior military officials criticize their civilian counterparts and superiors, even to the point of revealing privileged and even classified information, cannot be justified.

Yet civilian control remains very much alive and well. The many direct and indirect instruments of objective and subjective civilian control of the military suggest that the true issue is not control—defined as the government's ability to enforce its authority over the military—but rather political freedom of action. In virtually every sphere, civilian control over the military apparatus is decisive. All senior military officers serve at the pleasure of the President and can be removed, and indeed retired, without cause. Congress must approve all officer promotions and guards this prerogative jealously; even lateral appointments at the three- and four-star levels must be approved by the President and confirmed by Congress, and no officer at that level may retire in grade without separate approval by both branches of government. Operating budgets, the structure of military organizations, benefits, pay and allowances, and even the minutia of official travel and office furniture are determined by civilians. The reality of civilian control is confirmed not only by the many instances cited earlier where military recommendations were over-ruled. Not infrequently, military chiefs have been removed or replaced by the direct and indirect exercise of civilian authority.[40]

To be sure, the military as an institution enjoys some advantages. Large and well-trained staffs, extended tenure, bureaucratic expertise, cross-cutting relationships with industry, overt and covert relationships with congressional supporters, and stability during lengthy transitions between administrations give it a strong voice. But on the big issues of budget and force structure, social policy, and war and peace, the influence of senior military elites—absent powerful congressional and media support—is more limited than is often recognized.

If this thesis is correct, the instrumentalities and the efficacy of civilian control are not really at issue. As I have suggested, political freedom of action is the nub of the problem. Hampered by constitutionally separated powers which put the military in both the executive and legislative spheres, civilian elites face a dilemma. They can force the military to do their bidding, but they cannot always do so without paying a political price. Because society values the importance of independent, nonpoliticized military counsel, a civilian who publicly discounts that advice in an area presumed to require military expertise runs significant political risks.

40 In the decade of the 1990s, one Chief of Naval Operations was retired early following the Tailhook scandal. His successor committed suicide, troubled in part by persistent friction between senior naval officers and civilian defense officials he could not assuage. One Chief of Staff of the Air Force was relieved, and a Supreme Allied Commander in Europe and another Air Force Chief of Staff were retired early.

The opposition party will surely exploit any daylight between civilian and military leaders, particularly in wartime—hence the discernible trend in the modern era away from the Curtis LeMay's and Arleigh Burke's of yesteryear who brought powerful heroic personas and public reputations into the civil-military relationship.

It is, therefore, clear that much of the criticism directed at "political" soldiers is not completely genuine or authentic. Far from wanting politically passive soldiers, political leaders in both the legislative and executive branches consistently seek military affirmation and support for their programs and policies. The proof that truly apolitical soldiers are not really wanted is found in the pressures forced upon military elites to publicly support the policy choices of their civilian masters. A strict adherence to the apolitical model would require civilian superiors to solicit professional military advice when needed but not to involve the military either in the decision process or in the "marketing" process needed to bring the policy to fruition.

The practice, however, is altogether different. The military position of the Chairman of the Joint Chiefs of Staff (CJCS), the service chiefs, and the combatant commanders is always helpful in determining policy outcomes. The pressures visited upon military elites to support, or at least not publicly refute, the policy preferences of their civilian masters, especially in the executive branch, can be severe. Annually as part of the budget process, the service chiefs are called upon to testify to Congress and give their professional opinions about policy decisions affecting their service. Often, they are encouraged to publicly differ with civilian policy and program decisions they are known to privately question.[41]

This quandary, partly a function of the constitutional separation of powers and partly due to party politics, drives the CJCS and the chiefs to middle ground. Not wanting to publicly expose differences with the

41 The Army Chief of Staff's testimony on the Crusader artillery system cancellation in 2002 and postwar occupation policy in Iraq in 2003 are examples. See Robert Burns, "Rumsfeld set to change army leadership," *Associated Press*, April 26, 2003.

administration, yet bound by their confirmation commitments to render unvarnished professional military opinions to Congress, military elites routinely find themselves on the horns of a dilemma. These experiences, the bread and butter of military service at the highest levels, frequently produce exasperation and frustration. The consensus among civilian critics may be that the military dominates the policy process. But the view from the top of the military hierarchy is something quite different.

Conclusion

For military officers working at the politico-military interface, the problem of civil-military relations exists in its most acute form. There is, after all, no real issue between the polity as a whole and the military as an institution. Across the country, the armed forces are seen as organizations that work, providing genuine opportunities for minorities, consistent success on the battlefield and in civil support operations here at home, and power and prestige in support of American interests abroad. For most Americans, the military's direct role in the interagency process and in the making of national security policy is not only permissible but also it is essential to informed governance and a strong national defense.

The arguments advanced herein attempt to show that the dynamic tension which exists in civil-military relations today, while in many cases sub-optimal and unpleasant, is far from dangerous. Deeply rooted in a uniquely American system of separated powers, regulated by strong traditions of subordination to civilian authority, and enforced by a range of direct and indirect enforcement mechanisms, modern U.S. civil-military relations remain sound, enduring, and stable. The American people need fear no challenge to constitutional norms and institutions from a military which—however aggressive on the battlefield—remains faithful to its oath of service. Not least of the framer's achievements is the willing subordination of the soldiers of the state.

A Critique of the Maritime Strategy

In the 1980s, the Sea Services championed "The Maritime Strategy" as the answer to the Soviet's primacy on land, asserting that, as a "maritime power," the U.S. should rely on seapower as a primary, war-winning strategic approach. Echoes of the view that seapower in itself remains the most important capability in America's strategic arsenal are alive and well today. The Maritime Strategy was unquestionably effective in justifying and rationalizing a 600-ship Navy. But would it have worked?[1]

For most of the 1980s, naval leaders heralded the "Maritime Strategy" as the answer to the principal strategic challenges that faced the U.S. in the post-Vietnam era. As an historical case study, the Maritime Strategy still has much to teach about the endurance and persistence of service-specific approaches to national security, and about their limitations and dangers. As described elsewhere in this volume, the drive for service primacy and autonomy has not disappeared with the advent of Goldwater-Nichols. To an unusual degree, the Maritime Strategy advocated for a service solution to national security challenges. For this reason it merits close study even today.

In the 1980s, the primary conventional threat was seen to be a massive Soviet ground and air offensive in central Europe. Success or failure

1 This essay is based on an earlier work which appeared as "NATO's Northern Flank: A critique of the Maritime Strategy" in *Parameters*, June 1989.

hinged on the ability of ground forces coming from the continental U.S. to rapidly reinforce the central region. This in turn required secure sea lanes of communication across the Atlantic and a U.S. Navy optimized for anti-submarine warfare. Instead, the Maritime Strategy posited a very different approach.

Perceived American weakness and strategic incoherence undoubtedly contributed to the burst of Soviet adventurism that began in Angola in 1974 and culminated most spectacularly with the invasion of Afghanistan in 1979. Yet, the large military expenditures of the Reagan administration, intended to reverse the declining "correlation of forces" and reestablish U.S. military power as an effective instrument of policy, were not rigorously submitted to the discipline of an articulated and integrated conception of national military strategy. In the scramble for defense resources, each service advanced its own interests, but the sea services emerged clear winners, basing their requests for funds on the strategic necessity for a modernized, offensively-oriented, greatly-enhanced naval force structure. The Maritime Strategy focused on the Soviet Union and on naval and amphibious operations on NATO's Northern Flank as the answer to the U.S.S.R.'s military dominance in the Central Region. Navy strategy there was bold, aggressive, and dynamic in every way. It was also a likely prescription for failure.

To be sure, the Northern Flank was critically important to NATO. Control of the Norwegian Sea and the airfields in north Norway and Iceland could enable Soviet submarines and naval aviation to interdict NATO sea lines of communication (SLOCs) in time of war—potentially a war-winning strategy.[2] Apart from isolating Europe from the United States, control of the Scandinavian peninsula also would enable Soviet forces to exert pressure, in the form of air attacks, amphibious landings, and even conventional ground attacks, in support of operations in the Baltic, the Low Countries, and the North German Plain.

2 Kenneth Myers, "North Atlantic security: The forgotten flank?" *The Washington Papers*, 6(62), (London: Sage Publications, 1979), p. 64.

Naval and amphibious operations on the Northern Flank held center stage in the various formulations of the Maritime Strategy that evolved throughout the 1980s.[3] The strategy posited a phased naval campaign incorporating large carrier task forces, battleship surface action groups, amphibious warfare groups, and dozens of supporting surface and submarine combatants. Ground operations mounted from the sea were an important component of the strategy. Amphibious experts spoke of the "forcible entry of the 55,000 men of a Marine Amphibious Force"[4] where "amphibious forces can play a key role" in the "tense drama" of battle in the far north.[5]

In the initial stages of the conflict, called by maritime strategists the "deterrence or transition-to-war phase," naval forces would be marshaled in the form of the U.S. Striking Fleet Atlantic and deployed to the Norwegian Sea well before hostilities break out. This assumed that the problem of rapid response (always a concern with relatively slow-moving naval forces)[6] would be solved by an early political decision to commit military forces well before the crisis reached its flashpoint—a tenuous assumption at best.[7]

3 The most authoritative source document is *The Maritime Strategy*, published in January 1986 as a supplement to the January edition of US Naval Institute Proceedings. See also Michael A. Palmer, *Origins of the Maritime Strategy: American naval strategy in the first postwar decade* (Washington D.C.: Naval Historical Center, 1988).

4 James D. Watkins, *The Maritime Strategy*, p. 12. In 1988 the "amphibious" designation was replaced with "expeditionary;" the latter is used hereinafter.

5 P. X. Kelley and Hugh K. O'Donnell, Jr., "The amphibious warfare strategy," in *The Maritime Strategy*, p. 26.

6 Steaming time for the Strike Fleet Atlantic from the east coast to the North Atlantic was reckoned at seven days, not counting the time needed to collect the shipping and man the fleet (Watkins, p. 10). Deployment of a single Marine Expeditionary Force would take six weeks and require virtually every unit of amphibious assault shipping in existence. Michael Leonard, "Planning reinforcements: An American perspective," in *Deterrence and Defense in the North*, Johan Jurgen Holst (Ed.), (Oslo: Norwegian University Press, 1985), p. 164.

7 The decision to commit large naval forces early, the hinge on which the Maritime Strategy turned, is a subject of some confusion. "All of our war games, all of our exercises . . . indicate that in fact we will not make the political decision to move early." Admiral James Watkins, speaking in testimony before the Senate Armed Services Committee in 1984, cited by Jack Beatty, "In harm's way," *The Atlantic Monthly*, May 1987, p. 37.

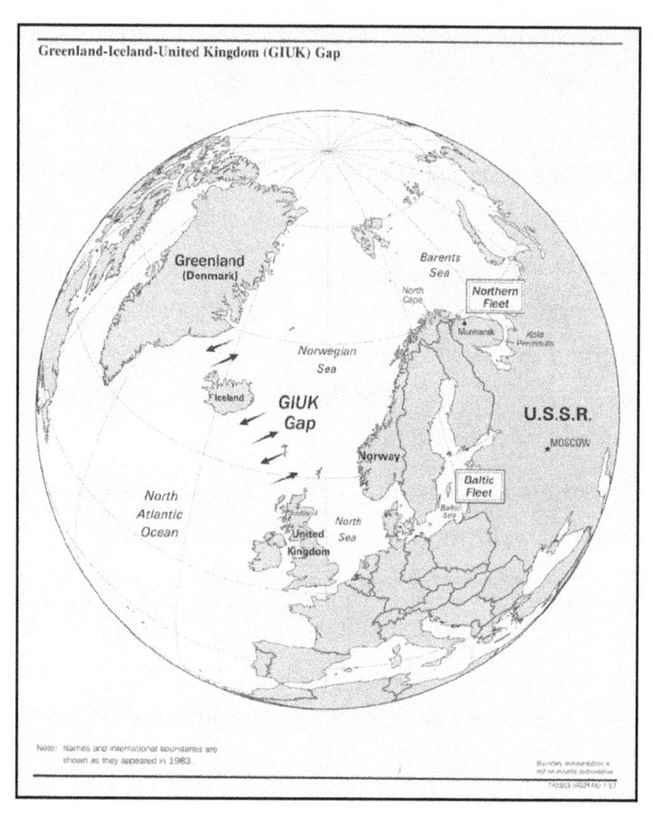

Figure 1: Greenland-Iceland-United Kingdom (GIUK) Gap
Source: "GIUK Gap," *Wikipedia*, 2019.

A Marine Expeditionary Brigade (MEB) was also slated to reinforce north Norway in the early stages of this transitional phase. However, its prepositioned materiel, unlike the prepositioned stocks of full divisional sets of equipment in Europe, consisted mostly of trucks and howitzers. Its helicopters, fixed-wing aircraft, tanks, other heavy support equipment, and many of its tactical vehicles would still come from the continental United States.[8] Furthermore, the MEB arrival airfields and equipment storage sites were (and are today) located in Trondheim in central Norway, hundreds of kilometers from the intended area of operations in Troms county. Finally, the addition of the three maneuver battalions

8 Harry J. Stephan, "USMC to bolster Norwegian flank: 4th MAB ready to deploy in a crisis," *Armed Forces Journal International*, August 1987, p. 34.

of the MEB would not materially alter the balance of forces in north Norway. Although the MEB might serve as a significant indication of allied resolve, its presence would fall well short of being enough to secure north Norway from ground attack by the Soviet army.

Phase Two of the Maritime Strategy was described as "seizing the initiative as far forward as possible".[9] This was amplified to mean that U.S. attack submarines would "roll back" successive antisubmarine barriers in the Norwegian and Barents Seas in order to scatter Soviet surface shipping and screening submarines. The ultimate objective was to attack and deplete the Soviet ballistic missile submarine (SSBN) force in its protected northern sanctuaries.[10] In theory, the antisubmarine campaign would contribute to SLOC defense by engaging Soviet attack submarines inside the Greenland-Iceland-Norway gap and alter the overall strategic balance by paring down Soviet SSBN strength.

Anti-air and anti-surface warfare were intended to complement the antisubmarine effort both by providing fleet defense and by destroying enemy forces in meeting engagements at sea. Implicit in this formulation was forward deployment of carrier groups inside the Norwegian Sea to conduct "aggressive, sustained forward operations."[11] MEB landings in the North Cape area were contemplated in Phase Two as an essential adjunct to the overall maritime objective of wresting the initiative from Soviet forces in the region.

These were ambitious objectives. The prospects for successfully achieving them were clouded by several factors, including the capabilities of Soviet land-based airpower, the effectiveness of the Soviet antisubmarine campaign, the success or failure of the territorial defense of

9 Also described as "establishment of sea control in key maritime areas as far forward as possible." The Norwegian Sea was specifically mentioned as an operational arena for carrier task forces in this phase. See Linton F. Brooks, "Naval power and national security: The case for the Maritime Strategy," *International Security, 11*, Fall 1986, p. 65.

10 See D. B. Rivkin, "No bastions for the bear," *US Naval institute Proceedings*, April 1984, pp. 36–43.

11 Watkins, p. 12.

north Norway and Iceland, and, most importantly, by the overall balance of forces.

At sea in the open ocean, carrier task forces are primarily concerned with defense against cruise-missiles and attack submarines. Screening vessels and the task force air umbrella virtually nullify surface threats. Within range of land-based aircraft, the threat level greatly increases. Even in the North Atlantic, however, the threat could be managed. At the distances involved, the Backfire and Bear bombers of 1980s Soviet naval aviation operated at the limit of their endurance. They could be tracked and engaged coming and going, providing plenty of reaction time to generate air defense. Operating inside the Greenland-Iceland-Norway gap, however, they posed a dramatically increased threat to carrier forces. Medium-range aircraft now became threats; so did land-based missiles and even tactical aircraft. Reaction times were slashed, the loiter time for enemy air threats went up, and the mere survivability of the carrier group, not its offensive function, became the priority.

A key premise was also that U.S. naval forces could defeat the Soviet submarine threat in the Norwegian and Barents seas. While the U.S. and NATO arguably enjoyed a qualitative edge, it is difficult to demonstrate that this offset the Northern Fleet's advantage in attack submarines. Designed and trained primarily for submarine and antisubmarine warfare, not blue water sea control,[12] the Northern Fleet possessed strong advantages in its own backyard. Besides its thirty-nine SSBNs (which did have organic self-defense measures, including torpedoes for protection against attack submarines), the Soviet Northern Fleet included 119 attack and cruise-missile submarines,[13] more than the entire complement of attack submarines in the entire U.S. Navy.[14] Pacific deployments, routine maintenance, and barrier operations to protect trans-Atlantic shipping

12 John Mearsheimer, "A strategic misstep," *International Security, 11* (Fall 1986), p. 37.

13 These forces are supported by 145 antisubmarine-warfare aircraft and seventy-one principal surface combatants, almost all of which are configured for antisubmarine operations. *The Military Balance, 1988–1989* (London: International Institute for Strategic Studies, 1988), p. 41.

14 Ibid.

would leave perhaps thirty U.S. attack submarines to execute the strategy, at odds of roughly one-to-five.[15]

The problem is magnified by the fact that the Soviet SSBN force could strike the continental United States by launching its MIRVed[16] nuclear missiles from underneath pack ice in the Arctic Ocean.[17] At the first sign of trouble, they could deploy rapidly under the ice, there to lie motionless and virtually undetectable.[18] Available NATO attack submarines,[19] operating in the ice without most of the complementary antisubmarine-warfare systems that normally aid them, would then have to discriminate between the target SSBNs and the Soviet attack submarines escorting them. Even if only ten percent of the SSBN force survives, its remaining counter value-targeted missiles could devastate every large metropolitan city in the United States.[20] In short, the belief that NATO naval forces could carry off such an ambitious program seems to presuppose an absolute technological and human superiority that is more optimistic than either reasoned judgment or historical experience can support. Physical occupation of north Norway and Iceland by the Soviets, together with Jan Mayen and Bear Islands and the Svalbard archipelago, were essential for effective prosecution of the Soviet antisubmarine-warfare effort, and, indeed, of their effort to control the Greenland-Iceland-Norway gap itself. Primarily for this reason, many experts considered these areas to be

15 Mearsheimer, p. 37

16 Multiple Independently-targeted Reentry Vehicle. This technology allows a single missile to deliver a number of warheads to different targets.

17 By the end of the decade, the mainstays of the Soviet SSBN force, the Delta III and Typhoon boats, would be armed with SSN-20 (8,300 km range, six to nine reentry vehicles) and SSN-23 (8,300 km range, ten reentry vehicles) submarine-launched ballistic missiles. See Edward B. Atkeson, "Fighting submarines under the ice," *US Naval Institute Proceedings*, September 1987, p. 82.

18 Ibid., p. 83.

19 It is important to include the submarine forces of the allied navies in the balance of forces in the north. Many of these, however, are diesel boats configured for coastal anti-shipping operations. It is unclear to what degree British, Norwegian, and German submarines would participate in anti-SSBN operations north of the Arctic circle. Their presence, in any case, represented only a marginal increase in the force of attack submarines, probably insufficient to change the equation much one way or the other.

20 Ibid., p. 82.

already "behind" Soviet lines from the psychological perspective of Soviet planners. In peacetime, these areas provided key facilities for NATO maritime patrol aircraft; seabed sensors; navigational, target acquisition, and communications systems; and naval bases. Without question, the Soviets could and would seize them at the outset.

Other issues also complicated the maritime thesis. The rationale for automatically attacking the Soviets's only protected second-strike nuclear deterrent is unclear, since Soviet land-based first-strike systems were not targeted for immediate destruction in the absence of a nuclear exchange.[21] This approach might inadvertently place Soviet decision makers in a "shoot'em or lose'em" position, with momentous consequences.[22]

The final phase of the Maritime Strategy was labeled "war termination," characterized by the exertion of global pressure against the Soviet Union; the total destruction of the Soviet navy; supporting the land battle by preventing redeployment of enemy forces, insuring NATO resupply and sustainment, and directly applying carrier and amphibious forces; and termination of the conflict through direct attack against the Soviet homeland or altering the nuclear correlation of forces.

In theory, the strategic application of maritime forces in the north during this phase would take place only after the Soviet Northern Fleet has been substantially eliminated. Direct attack against the Kola infrastructure and the Soviet homeland itself was postulated, implying that Soviet naval forces that might prevent closure to within striking ranges of these targets had been destroyed. At this point, the Navy's power projection capabilities were expected to take effect and the combined weight of carrier aviation, cruise missiles, surface action groups, and amphibious

21 Some analysts recognized this inconsistency but dismiss it with vague references to "competitive substrategies" such as the use of stealth platforms to strike at Soviet land-based ICBMs. There was, however, no publicly expressed intent to automatically target these systems for destruction while the conflict remains conventional; the existence of a theoretical capability to do so in no way implies such intent. See F. J. West, "The Maritime Strategy: The next step," *US Naval Institute Proceedings*, January 1987, p. 41.

22 Lawrence Freedman, "Concluding remarks," in *Britain and NATO's Northern Flank*, Geoffrey Till (Ed.), (London: The MacMillan Press, 1988), p. 18.

landings brought to bear to force a decision on land. For the Marines, the likely objective was North Cape.[23] For the Navy, the objective was the huge and costly naval and air facilities of the Kola Peninsula—the home anchorage of the Soviet Northern Fleet.[24]

Though serious criticism of U.S. capability to conduct large-scale opposed landings from the sea against Soviet forces was only sparsely represented in the literature, the requirement for almost total sea and air dominance gravely complicated the operational feasibility of such landings.[25] The Marine Expeditionary Force, coming in amphibious warfare groups observable by satellite reconnaissance, could not contend with active opposition from land, sea, and air during its vulnerable transition from ship to shore. The assumptions supporting a large amphibious assault in this scenario were formidable: the virtual destruction of the Soviet Northern Fleet and its supporting air component; no diversion of Marine combat or assault shipping assets to other theaters; and the arrival of the amphibious assault groups from the United States intact.

The validity of the concept of direct attack against Soviet targets ashore, a fundamental part of the philosophy behind the Maritime Strategy, was, therefore, open to question, even in the unlikely event that all of these enabling assumptions were realized. Publicists often described Marine units as though the entire complement functioned as combatants. The ringing claim—the "forcible entry of the 55,000 men of the MAF"—is representative. A prominent admiral writing at the time boasted that a four-carrier battle force and a Marine Expeditionary Force could deploy over 50,000 combatants and 900 tactical aircraft anywhere between Norway and Turkey in a matter of days.[26]

23 Kelley and O'Donnell, p. 26.

24 "The most valuable piece of real estate on earth," John Lehman, quoted by Michael Gettler, "Lehman sees Norwegian Seas as key to Soviet naval strategy," *The Washington Post*, December 29, 1982, p. 4.

25 Mearsheimer, p. 25.

26 Dennis Blair, "The strategic significance of maritime theaters," *US Naval War College Review, 41* (Summer 1988), p. 38.

Statements like these were at best misleading.[27] In point of fact, the Marine Expeditionary Force included a very large air wing and forward service support group, as well as many service support troops within the division itself. In terms of maneuver units—the fighting core of the force which seizes and holds ground—the Marine Expeditionary Force fielded only nine light infantry battalions and a single tank battalion.[28] Fifty-seven percent of the "tactical aircraft" referred to were not combatants at all, while only a third could be used against targets ashore.[29]

In contrast, Soviet forces earmarked for the region, the "Northwest TVD" in Soviet parlance, consisted of an airborne division, an air assault brigade, a naval infantry brigade, a naval *spetznaz* brigade, and eleven heavy divisions, including two border divisions equipped with over-snow vehicles and trained for amphibious operations.[30] After mobilization, the Norwegians could deploy four (mostly reserve) brigades in Finnmark and Troms counties, assisted by several allied battalions from the Allied Command Europe Mobile Force and the United Kingdom/Netherlands Amphibious Force, if available.[31] In short, without substantial ground forces to follow up the assault landing of the Marine Expeditionary Force,[32] the prospects for success, even in a best-case scenario, were low.

27 The U.S. Marine Corps is justifiably proud of the combat skills of its support troops, but these skills translate into a more effective capability for rear area protection and self-defense, not a capability to use support units in maneuver roles.

28 Approximately 7,000 men, with perhaps another 3,000 in combat support roles such as reconnaissance, artillery, air defense, and combat engineer.

29 The Marine Expeditionary Force standard air wing fielded 333 aircraft, of which 152 can be classified as combatants (AV-8, F/A- I8, A-6E, and AH-1). The standard complement for the carrier air wing varied at the time but normally included only forty strike aircraft (A-6E and F/A-18). The rest were designed for fleet air defense, electronic warfare, search and rescue, antisubmarine warfare, or refueling and could not be used to attack shore targets. See *Sea power, 31* (January 1988).

30 "The border divisions and all the specialized units listed are category I." *Deterrence and Defense in the North*, p. 103. "It is estimated that there is enough commercial shipping in the Kola ports at any time to lift and deploy two heavy divisions as a landing force in support of amphibious operations in north Norway." Edward Fursden, "The Kola," *Army Quarterly and Defence Journal, 118* (April 1988), p. 179.

31 II Marine Expeditionary Force, the United Kingdom/Netherlands Landing Force, and Allied Command Europe Mobile Force were not solely dedicated to north Norway. Iceland, Denmark, Turkey, or the Mediterranean represented other contingency areas for these forces. Norwegian force structure is discussed by Tonne Huitfeld in *Deterrence and Defense in the North*, p. 175.

32 A conservative estimate of the forces needed to follow up a Marine Expeditionary Force

The modest threat posed by the activities of an amphibious contingent operating in the far north was unlikely at best to cause the Soviets to redirect meaningful resources away from the Central Front—a stated objective of the final phase of the Maritime Strategy.

In this final phase of the struggle, some of the NATO naval forces in the region, primarily battleship surface action groups and perhaps a carrier, would support amphibious operations. The bulk of the fleet, however, would concentrate on the destruction of the remnants of the Soviet Northern Fleet and move in for direct attacks against the Soviet homeland itself. The richest prize was the Kola Peninsula, at the time probably the most heavily militarized area in the world.

The general intent was to move into the Barents Sea and launch concerted air and missile strikes against the Kola. Navy F/A-18 Hornets and A-6E Intruders would spearhead these attacks with precision-guided munitions and cruise missiles, backed up by cruise missile strikes from submarines and surface platforms.[33] Maritime Strategy proponents insisted that, by exerting such "direct pressure" against sensitive flanks, "fear, uncertainty, and paralysis" could be induced in the mind of the aggressor, thereby relieving pressure on the Central Front in some undefined way. [34] At the outset, the naval air strike components of this proposed force could muster, assuming total availability of pilots and aircraft, perhaps 160 aircraft;[35] remaining aircraft were dedicated to protecting the fleet. In fact, virtually the entire carrier task force, with its Aegis cruisers, frigates, destroyers, escort submarines, and fast support ships, existed to support the 40-strike aircraft on each carrier. At night or in bad weather, only the A-6E Intruders had an advanced all-weather/night-attack capability.

landing is an Army corps. Mearsheimer estimates that at least five heavy divisions would be needed. Cited in Beatty, p. 46.

33 "The newest Seawolf-class attack submarines, when deployed, will carry 50 vertical cruise missile launchers, up from the ten now fitted on front-line Los Angeles-class attack submarines." Lawrence R. Nilssen, "Nordic NATO in transition," *Air Power Journal*, Summer 1988, p. 71.

34 Kelley and O'Donnell, p, 26.

35 A more realistic estimate for sustained combat operations is 100–120 aircraft due to attrition, pilot casualties, maintenance down-time, and fatigue,

To face this threat, the Kola was defended by more than 225 modern Soviet air defense fighters of all types, dispersed over sixteen major airfields and many smaller ones. These were supplemented by the land-based bombers and fighters of Soviet naval aviation based there. In wartime, and particularly in response to a direct threat to the Kola, these forces would be stiffened considerably.[36] Closing at top speed, the carriers of the striking fleet would come under attack but much stronger forces for two days without being able to respond with their own strike aircraft.[37] The Soviet air defense effort was further supported by more than 100 surface-to-air missile installations, as well as numerous radar tracking sites for target acquisition and ground control of defending aircraft.[38]

The underlying premise here—that sea-based airpower could compete with and overcome land-based airpower—fails to pass any realistic feasibility analysis.[39] Though the carriers might reposition, they could not hide from Soviet tactical and satellite reconnaissance indefinitely. They would be especially vulnerable while launching and recovering aircraft, while the strain of round-the-clock air operations for fleet air defense would mount quickly. Soviet ground-based air defense squadrons could easily out-sortie, out-resupply, and out-reinforce the carrier air wings because of the intrinsic advantages of land bases and by simply flying in personnel and spares from the interior. The Kola Peninsula was a very large, very hard target, well protected and defended, and invulnerable to destruction from the air by anything less than a sustained, heavily-weighted aerial campaign. At the time, in fact, it was the most heavily-defended real estate in the world.

What were the potential consequences of a mismatch between sea-based and land-based airpower in the north? The first and most obvious

36 It was estimated that the number of operational aircraft in the Kola could be doubled without notice (Fursden, p. 79).

37 George Thibault, cited in Beatty, p. 44.

38 *The Military Balance*, 1988–1989, p. 40.

39 "The Navy, if it is serious about launching a large-scale air offensive against the Kola, would surely have to augment its forces with significant numbers of land-based fighters and bombers." Mearsheimer, p. 38.

was the loss of the American carriers. A defeat of this kind would be a serious material blow and a spectacular psychological coup for the Soviets. Aircraft carriers have come to be symbols of almost mythic proportion; much more than mere floating airfields, they symbolize national prestige, military potency, and strategic reach. To be effective, they must be put at risk, but only where the potential returns justify it. The gamble here was a poor one, and no claims of Soviet submarines sunk, enemy aircraft downed, or airfields knocked out would soften the impact of such a severe blow.

Another potential problem was the unintended threat of nuclear escalation. The primary naval weapons for strikes against shore targets, the strike aircraft and cruise missiles, were dual-capable.[40] Whether they were used in a conventional or nuclear mode could not be determined prior to use. The Tomahawk cruise missile in particular carried a high potential for destabilization because of its range, accuracy, and survivability. Would Soviet commanders initiate nuclear warfare at sea only *after* suffering a sea-launched nuclear attack themselves? After detecting inbound missiles and aircraft, they might well have responded with nuclear weapons while they still had the capacity to do so.

Other important issues intrude as well. Because of such large U.S. forces deployed in an offensive sea control mode inside the Greenland-Iceland-Norway gap, the Atlantic SLOCs would be denuded of most of the available at-sea air cover and many of the surface and subsurface antisubmarine-warfare platforms needed to insure resupply for Europe. Sea control of the Atlantic was an essential pre-condition for victory on land in Europe. Maritime strategists played down this requirement by insisting that SLOC interdiction was not a real Soviet priority (despite the fact that the Soviet Union touted the submarine as the primary conventional

40 "Maritime forces can influence [the nuclear correlation of forces] both by destroying Soviet SSBNs and by improving our own nuclear posture through the deployment of carriers and Tomahawk platforms around the periphery of the Soviet Union." Watkins, cited in Ronald O'Rourke, "Nuclear escalation, strategic anti-submarine warfare and the navy's forward Maritime Strategy," *Congressional Research Service Report* No. 87-138 F, 27 February 1987, p. 8.

and nuclear striking arm of its navy)[41] and by criticizing SLOC defense strategies as passive approaches which left the initiative to the enemy.[42]

Further evidence that the Soviets prioritized sea denial in the North Atlantic is provided by a simple analysis of Warsaw Pact naval force structure:

> [T]he Pact's major sea goal will be to deny the Allies the use of the sea lanes. This consideration is reflected in the composition of Pact naval forces. The Warsaw Pact has 145 long-range attack submarines, NATO some 68. Since 1983 the Soviets have introduced four new classes of nuclear-powered attack submarines, further emphasizing their commitment to large numbers of sophisticated sea denial units.[43]

In short, the evidence that Soviet strategists had abandoned any hopes of strangling the Alliance with a vigorous sea-denial/SLOC-interdiction campaign was absent. If merely defending ballistic missile submarine sanctuaries had become the raison d'etre of the Soviet Northern Fleet, prudence would dictate the deployment in large numbers of cheaper, quieter, modern versions of the diesel-electric SS boats. This did not happen. Soviet attack boats, if not as quiet as U.S. versions, were fast, deep diving, and numerous. The rationale for pressing an intensive surface and sub-surface campaign in northern waters, predicated on an assumption that SSBN protection was the top priority for Soviet attack subs, was not compelling.

Heavy U.S. losses in the Norwegian Sea, which could not be ruled out even by the most optimistic advocates of the Maritime Strategy, meant correspondingly fewer assets to fight the next "Battle of the Atlantic"—a

41 Victor Suvorov, *Inside the Soviet army*, 2nd edition (New York: Berkeley Publishing Co., 1984), p. 102.

42 "Sea lane interdiction is a lower priority than protecting sea-based strategic forces or homeland defense." Brooks, p. 71.

43 Karsten Voight, "Alliance security," in *The State of the Alliance 1987–1988* (Boulder Colorado: Westview Press, 1988), p. 71.

battle that literally would spell life and death for those fighting the decisive land engagements in central Europe. And the balance of forces in the north, coupled with the fact that the Soviets were rapidly deploying another secure second-strike deterrent in the form of land-based mobile ICBM launchers,[44] suggests that many modern attack and cruise missile submarines would be available to conduct sea-denial/SLOC-interdiction missions in the open ocean. In the final analysis, there was no such thing as a naval strategy that realistically accommodated both SLOC defense and early forward deployment of the bulk of the Atlantic Fleet.

From a joint perspective, the assumptions and priorities that lay behind the Maritime Strategy emphasized naval solutions to problems with important and substantive land and airpower dimensions as well. Concerned observers noted that maritime forces and land forces are effective against each other only at the margins where they meet, not well inside their respective mediums. As one analyst pointed out:

> The U.S. Maritime Strategy and the U.S. Navy can contribute to a denial of Soviet objectives on land in Europe, but they cannot substitute for conventional force or doctrinal deficiencies relative to the land battle. As Admiral Isaac Kidd is reported to have said, someone must take the land and say "this belongs to me." Navies cannot do that, whatever else they may do. It would be wrong and misguided in the extreme for U.S. maritime strategists to offer their forces as substitutes for increased supplies of ammunition or additional operational reserves in Europe.[45]

The naval solution, emphasizing naval airpower and amphibious forces that can be projected ashore, was a partial one at best because

44 The Congressional Research Service estimated that the Soviets would deploy between 1,950 and 2,850 mobile ICBM warheads by the early 1990s. O'Rourke, p. 15.

45 Stephen J. Cimbala, *Improving extended deterrence in Europe* (Lexington, MS: Lexington Books, 1987), p. 173.

naval aviation is both relatively short-ranged and vulnerable and because Marine forces are not capable of sustained land combat against large, armored forces.

The Maritime Strategy was also suspect because it was designed to generate heavy requirements in terms of ships, men, and materiel. It demanded resources because it was highly resource-consumptive; it proposed to engage enemy forces as aggressively, as far forward, and as directly as possible, against the hardest and strongest targets, conceding important geostrategic and force-ratio advantages to the Soviets. Nowhere on earth was the combination of Soviet naval strength and U.S. vulnerability so telling as in the Norwegian and Barents Seas. Yet the Maritime Strategy purported to play to this strength virtually from the opening bell.

The contradictions found in the Maritime Strategy proved so glaring, so parochial and service specific, that many inside and outside the Navy viewed the strategy more as a public narrative, intended to justify a larger budget and a bigger battle fleet. In this, the Navy largely succeeded. Today, the U.S. battle fleet is larger than the next thirteen navies combined, of which eleven are allies, while fielding more nuclear attack submarines and fleet carriers than the rest of the world combined.[46] Nor has the idea itself been subsumed by joint doctrine. Official Navy documents continue to assert "seapower has been and will continue to be the critical foundation of national power and prosperity and international prestige for the United States of America . . . assuring freedom of action in any domain—sea, air, land, space or cyber."[47]

As the most formidable maritime power on earth, with self-contained and powerful naval, amphibious, and air forces all its own, the Department of the Navy is unquestionably a vital and indispensable element of national security. Its role in policing the global commons,

46 Robert Gates, remarks to the Navy League, May 3, 2010.

47 *A cooperative strategy for 21st century Seapower,* (Washington, D.C.: Department of the Navy, 2016).

securing the sea lanes, ensuring access, and enabling power projection cannot be overstated. Naval proponents are correct when they opine that America "is a maritime power." Yet, there is much to suggest that the historical conundrum of seapower vs. land power still contributes to a dislocation of strategic thought in America. Hard experience in multiple campaigns teaches that seapower is necessary but not sufficient in campaigns fought on the land. In truth, the question has never been one of continental vs. maritime strategies. Modern war transcends such categories. The joint application of balanced land, sea, air, space, and cyber forces capable of achieving their strategic objectives within the context of an intelligent national security strategy is the ultimate goal. The issues raised herein suggest that we can do better than the narrow prescriptions called for in the Maritime Strategy.

8 The Role of Airpower in Modern War

Almost a century ago, Guilio Douhet argued that airpower alone held the key to victory in war. The doctrine of strategic bombardment—enshrined in Air Force doctrine today as "Strategic Attack"—remains central to Air Force thinking about how to fight and win wars. Airpower advocates to the present day continue to argue that airpower is the decisive factor in modern war and that budget and strategy decisions should reflect that primacy. In this essay, the author reviews the historical record and concludes that, while a critical tool in America's strategic inventory, airpower has repeatedly fallen short of its promise.[1]

Since the Second World War, airpower has been the crown jewel in America's strategic inventory.[2] Since 1944, American dominance in the air has been supreme, yielding absolute superiority in the air in every conflict. This success has led to a body of thought about airpower, based on deep roots, that challenges true jointness. Many airmen believe that airpower alone can deliver strategic success and that our strategic

1 An abridged version of this essay appeared as "Airpower in American wars," *Survival, 58*, June 2016.

2 Though official U.S. Air Force doctrine has used the term "aerospace power" since 1992, "the most fundamental operational and strategic characteristics of air power are almost wholly absent in the space arena." See Karl P. Mueller, *Airpower*, The RAND Corporation, 2010, p. 16. This paper focuses on the employment of aircraft, manned and unmanned, at altitudes low enough to sustain aerial flight, hence the use of the more traditional "airpower." Aerial delivery of nuclear weapons also occupies a decidedly distinct and separate space and lies outside this discussion.

and resource decisions should reflect that primacy. Historically, this narrative has fallen short of its promise, while the drive to centralize control of airpower has created tensions within the joint community that persist today.

Giulio Douhet's *The Command of the Air*, published in 1921, established a theory of the dominance of airpower over armies and navies that endures to this day.[3] Building on Douhet's theories, the belief that wars can be won solely or primarily from the air is canonical for many airmen.

> Most airmen subscribe to the notion that superior technology can produce economy of force in any context, echoing early airpower theorists who believed mastery of airpower technology was the key to avoiding massive, force-on-force struggles.[4] These ideas have been central to the Air Force since its birth as a service, and are deeply ingrained. WWII was a proving ground for key airpower tenets while also providing the vehicle by which the USAF traveled to institutional independence.[5] Having earned its stripes, the young Air Force found itself constantly fighting for relevance, and this constant fight transformed core airpower ideas into indivisible beliefs. The net result over time has been an Air Force perspective zealously wedded to technology, offensive operations against vital centers, and a drive for decisiveness.[6]

Airpower, of course, is much more than the combat aircraft in the inventory. It includes the industrial and technology base; pilots and

3 Guilio Douhet, *The command of the air*, reprinted by the Air Force History and Museums Program, Washington D.C., 1998.

4 David R. Mets, *The air campaign: John Warden and the classical airpower theorists* (Montgomery Alabama: Air University Press, 1998), p. 74.

5 Richard J. Overy, *The air war: 1939–1945* (Washington, D.C.: Potomac Books, 2005), xi, p. 10; Carl H. Builder, *The Icarus syndrome: The role of air power theory in the evolution and fate of the US air force* (New Brunswick, CT: Transaction Publishers, 2003), pp. 133–136.

6 Anthony Carr, "America's conditional advantage: Airpower, Counterinsurgency and the theory of John Warden," *School of Advanced Air and Space Studies*, June 2009, p. 25.

aircrew and the training base which produces them; the ISR, command and control, and tanker communities which enable the combat mission; the strategic and tactical airlift communities which are so crucial to American power projection; air intelligence, logistics, maintenance and targeting communities; and much else. The airpower provided by partners and allies is also a critical part of the equation and one often overlooked in airpower discussions. Airpower operates to greatest effect when fully synchronized with the Joint Force and the interagency. Without question, airpower has played a central and indispensable role in every American conflict since WWII.

The belief in airpower dominance carries a corollary: that airpower is by its very nature a strategic instrument in a way that land power and seapower are not. Airpower theorists have long contended that wars can be won from the air alone by attacking "strategic" targets (such as an adversary's industry or means of command) far beyond the battlefield.[7] The notion that airpower is inherently strategic carries with it the implication that operational and tactical uses are less legitimate. From the beginning, this was a necessary argument if air forces were ever to achieve a separate and equal status with sister services.[8] Though resisted by more traditional and conservative armies and navies, airpower offered dazzling advantages that were spelled out as early as the 1920s. After the horror of the trenches, the prospect of rapid, decisive victory without huge casualties and expenditures proved almost irresistible.[9]

7 Channeling Douhet, U.S. airpower theorists in the 1930s wrote that aerial attack of transportation, power, and water systems would "quickly and efficiently destroy the people's will to resist." Howard D. Belote, "Warden and the Air Corps Tactical School," *Airpower Journal*, Fall 1999, p. 42.

8 "The very creation of the Air Force as an independent service grew out of the U.S. effort in World War II to mount a decisive air campaign to destroy the enemy's capability to wage war and to decide the conflict independent of military operations on land and sea." Richard Kohn, "U.S. strategic airpower 1948–1962," *International Security*, Spring 1988, p. 79.

9 "... those who daringly take to the new road first will enjoy the incalculable advantages of the new means of war over the old ... those who are ready first not only will win quickly, but will win with the fewest sacrifices and the minimum expenditure of means." Douhet, p. 30.

Airpower in World War II

The Second World War provided the first opportunity to put these ideas to the test. Enormous resources were poured into the British and American strategic bombing offensives in WWII. During the war, Allied aircrew losses were extraordinary,[10] while civilian casualties and destruction were enormous. The German air force was swept from the skies at great cost, while the diversion of German war industry for air defense was highly significant. This effort had results; few historians attempt to argue that strategic bombing failed to exercise a major effect on the outcome of the war.[11]

Nevertheless, most objective scholarship concluded that airpower fell short of the more extreme claims made on its behalf. By itself, it had not been able to win the war. In conjunction with American and Allied land and seapower, airpower had played a critical role in securing victory. German oil supplies were badly hit, submarine manufacturing was curtailed, and ammunition production fell sharply off in the last year of the war. Below the strategic level, German and Italian forces faced great difficulties in movement of mechanized forces in the face of allied tactical airpower. On the other side of the ledger, German tank and aircraft production peaked in 1944, the heavy emphasis on targeting the ball bearing industry produced only limited results, and German consumer goods and food supplies remained adequate throughout the war. Germany was able to maintain some 300 divisions in the field right up until the end of the war. Heavy, even saturation bombing of urban centers day and night did not lead to regime collapse.

In the Pacific, strategic bombing played a different role. Only 160,800 tons of bombs were dropped on the home islands (compared to 2,700,000 tons dropped over Europe), only 10% of which hit near the target until the Army Air Forces went to low level, nighttime, area attacks

10 British and American losses in the air were almost identical: 79,281 vs 79,265. See *The United States strategic bombing surveys*, reprinted by the Air University Press, Maxwell Air Force Base: Alabama, October 1987.

11 U.S. aircraft losses during WWII totaled 95,000, including 52,951 operational losses. See John Ellis, *World War II—A Statistical Survey*, Facts on File, October 1993.

against urban areas. In 1945, Japanese production dropped below 50% of pre-war totals and food supplies were reduced. Against the Imperial Japanese Navy and the Japanese merchant fleet, only a small fraction were destroyed by land based bombers, with most sunk by tactical Navy and Marine aircraft and submarines.[12]

In both theaters, the theory of strategic bombing embraced in WWII centered on bombing cities, in line with Douhet's proposition that destroying an enemy's will to resist through deliberate targeting of civilian populations was the quickest and surest path to victory. His assertion that "there will be no distinction any longer between soldiers and civilians" was put into effect with a vengeance. Called "morale bombing" or "de-housing" by the British, and influenced in part by the relative inaccuracy of high-altitude bombing, this program eventually saw the deliberate use of incendiaries to effect even greater destruction.[13] By war's end, the civilian death toll stood in the hundreds of thousands. While both the Germans and Japanese bombed civilian targets during the war, their lack of true strategic bombers like the B17 and B29 and their lack of industrialization relative to the U.S. meant that their efforts were far less casualty producing. While broadly supported at the time, the deliberate targeting of civilian populations from the air increasingly came to be seen after the war as morally repugnant.

The Allied use of airpower for tactical and operational uses (both combat and airlift) in WWII usually receives less attention but played a major role in every theater. Whether interdicting enemy forces behind the front, providing direct support to front line troops, or moving men and materiel around the world, these phases of the air war were carried out on a colossal scale as well. The abundance of resources made available to the

12 Four-star Admiral Raymond Spruance, the hero of Midway, "is reported by one reputable Navy historian to have made the flat statement 'the submarine beat Japan.'" Cited in Carl Builder, *The masks of war: American military styles in strategy and analysis* (Baltimore: The Johns Hopkins University Press, 1989), p. 77.

13 "[In WWII] only about 18% of U.S. bombs fell within 1000 feet of their targets." Robert Pape, "The true worth of airpower, *Foreign Affairs*, March/April 2004, p. 1.

air arms of the different services almost beggars belief; beginning with an inventory of 1,700 aircraft in 1939, U.S. industry produced 295,959 planes during the war, of which 200,443 were combat types.[14] This industrial backbone, plus a pilot training system second to none, gave America the edge in airpower by a wide margin.

Airpower in Korea

In the Korean conflict, policymakers and military commanders initially placed limits on bombing to avoid intentionally targeting civilian populations. However, battlefield reverses led to a reexamination of policy and, within six months, the U.S. adopted a relaxed policy, stretching the "definition of "military target" far beyond its original meaning."[15] Though exact numbers are not known, civilian casualties were very high. Given the moral debate which had emerged following WWII, military publicists and headquarters reports generally described air operations in terms of military targets, with civilian casualties addressed as unintentional collateral damage. This approach masked a use of airpower that approached war *à outrance*.

By 1951, air superiority was firmly established, and the entrance of Chinese forces into the war had been contained and the lines stabilized. The Far East Air Force Bomber Command now focused on interdicting North Korean and Chinese supply lines in Operation Strangle and Operation Saturate, hitting railway lines, bridges, and marshalling yards with "marginal success."[16] In 1952, in an effort to drive the enemy to the negotiating table, aerial bombardment was stepped up with large scale attacks on hydroelectric plants and logistics centers, in some cases with

14 Arthur Herman, *Freedom's forge: How American business produced victory in World War II* (New York: Random House, 2012), pp. 202–203.

15 Sarh Conway-Lanz, "America's ethics of bombing civilians after World War II: Massive casualties and the targeting of civilians in the Korean War," *The Asia Pacific Journal*, September 15, 2014.

16 James A. Grahn and Thomas P. Himes, Jr., "Air power in the Korean War," *Air Command* and Staff College, Maxwell Air Force Base, April, 1998, p. 20. A similar interdiction effort, also named STRANGLE, was attempted in Italy in 1943–1944 which "impaired but did not critically delete German access to fuel and ammunition." F. M. Sallagar, "Operation Strangle: A case study of tactical air interdiction," *The RAND Corporation*, February 1972, p. 9.

up to 500 aircraft (including U.S. Navy planes), later broadening the target list to focus on irrigation dams. This operational use of airpower undoubtedly pressured the enemy and materially assisted ground forces. True "strategic" bombing was not pursued because the obvious targets lay outside North Korea, in China, because the primitive nature of North Korean infrastructure did not lend itself to strategic targeting and because much of the strategic bombing force was held in reserve for possible use against the Soviet Union. Even so, the effort expended was enormous. In only three years, U.S. and UN forces flew more than a million sorties and delivered more than twice the tonnage of bombs dropped on Japan.[17] Estimates of North Korean deaths from air strikes approximate 20% of the population or 1–2 million, "the majority civilian."[18]

While the Air Force was unable to pursue true strategic bombing, its role as the principal airpower provider for a large-scale land campaign brought it into conflict with the Army over the proper use of airpower resources and over their command and control. Established as an independent service only three years before, the U.S. Air Force was keen to assert its primacy. Persistent disputes over these issues rose to the level of the UN Supreme Commander, General Mark Clark, and were eventually referred to Washington without resolution. The inherent speed and power of fixed wing aviation drove the most senior Army commanders to demand operational control of "tactical" Air Force assets for interdiction and CAS, as enjoyed by Marine commanders employing Marine air.[19] This tension would carry over into every future conflict.

17 According to the Air Force Historical Research Agency, the U.S. Air Force suffered 1,841 battle casualties and lost 1,466 aircraft in the Korean War.

18 Marilyn B. Young, *Bombing civilians: A twentieth century history*, Yuki Tanaka and Marilyn Young (Eds.) (New York: The New Press), 2009, p. 157. In retirement, former Air Force Chief of Staff General Curtis Lemay was quoted as saying "We burned down every town in North Korea." See *Strategic air warfare: An interview with Generals Curtis E. Lemay, Leon W. Johnson, David A. Burchinal and Jack J. Catton*, Richard Kohn (Ed.) (Washington, D.C.: Office of Air Force History, 1988).

19 The average response time for an Army request for CAS in Korea was forty to fifty minutes, but only ten to fifteen for the Marines, who owned their own tactical air and who employed different procedures. Bryan K. Luke, "Will close air support be where needed to support objective force operations in 2015?" *U.S. Army Command and General Staff College*, 2002, p. 20.

Airpower in Vietnam

Vietnam was the next major conflict and, as in Korea, airpower played a key role. Beginning in February 1965, Operation Rolling Thunder commenced, a "measured and limited" effort marked by escalating and deescalating bombing attacks against the "Democratic Republic of Vietnam" (DRV), designed to degrade North Vietnam's military effort in the south.[20] The Johnson administration ended Rolling Thunder in March 1968, and, for the next four years, strategic targets in the north were left alone. The "Vietnamization" policy adopted by the Nixon administration saw a return to large-scale bombing in Laos and Cambodia against communist staging areas as well as in North Vietnam. The breakdown of talks in 1972 spurred a resumption of strategic bombing with Operations Proud Deep and Linebacker I. Using new laser-guided bombs, U.S. airpower struck fuel stores, railroads, power plants, and other key infrastructure targets.[21] Linebacker I was judged to have accomplished more in four months than Rolling Thunder in four years, leading to a reopening of talks. When these broke down in December 1972, Operation Linebacker II delivered more ordnance against strategic targets in the north in twelve days than in the previous two years. Having expended most or all of its inventory of surface to air missiles, North Vietnam consented to the Paris Peace Accords, leading to the withdrawal of U.S. military forces from Vietnam.[22]

At the tactical level, after early disputes, the integration of air and ground operations improved over Korea, with 60% of Army general officers rating Air Force cooperation as "outstanding."[23] With few set-piece battles and a non-linear battlefield, "all air-ground attacks within South

20 Abigail Pfeiffer, "Airpower in Vietnam: A strategic bombing analysis," *Exploring Military History and Humanity*, September 15, 2011.

21 Gunter Lewy, *America in Vietnam* (Oxford UK: Oxford University Press, 1978), p. 410.

22 The U.S. Air Force suffered 2,586 fatalities in the Vietnam conflict and lost 2,251 aircraft, including 1,737 to hostile action. The U.S. Navy lost 859 aircraft, 530 in combat, and 441 aircrew killed or missing. The U.S. Marine Corps lost 193 fixed wing aircraft in combat and 270 helicopters. The U.S. Army lost 364 fixed wing aircraft (mostly OV1 and O-1 light observation planes) and 5,086 helicopters. Source: *The National Archives*, Electronic Records Reference Report.

23 Douglas Kinnard, *The war managers* (Hanover, NH: University Press of New England, 1977), cited in Luke, p. 32.

Vietnam were considered CAS." Army commanders were provided with tactical air control parties down to battalion level. Each corps or "Field Force" headquarters was equipped with a Direct Air Support Center to process requests. By 1965, with the introduction of major ground units, the air-ground request system in place more closely resembled the superbly integrated MAGTF. CAS sorties were normally airborne and available with a 20-minute response time, backed up by ground alert aircraft with a 40-minute response time. The ubiquitous use of airborne controllers (forward air controllers-airborne or "FAC-As") and dedicated CAS platforms, in particular the A1 Skyraider, proved extremely capable.

While ground commanders saw improvement over Korea, air commanders were less sanguine. The air war in Vietnam was far from centralized. While General Westmoreland, the Army commander of Military Assistance Command Vietnam (MACV), controlled the air war in South Vietnam through his Air Force deputy, the Commander U.S. Pacific Command directed the war over North Vietnam. Naval and Marine aviation remained under service control, with B52 strategic bombers controlled by the Strategic Air Command. This dispersion of the air effort led Westmoreland in 1968 to implement a "single manager for air," placing Marine fighters under control of the Commander 7th Air Force, Lieutenant General Momyer. Though ultimately imposed, this move provoked stout opposition from the Marine Corps and an inter-service controversy that ultimately reached the secretary of defense.[24]

The failure of the American effort in Vietnam cannot be ascribed to a failure of airpower. U.S. airpower delivered eight million tons of bombs against communist targets, "twice the amount dropped in WWII by all the belligerent nations combined." Nineteen million gallons of defoliant were dropped, destroying crops over 20% of South Vietnam or six million acres. In Laos alone, "three million tons were dropped, exceeding

24 See Willard J. Webb, "The single manager for air in Vietnam," *Joint Force Quarterly*, Winter 1993–1994.

the total for Germany and Japan by both the U.S. and Great Britain."[25] However, political restrictions constrained air commanders from attacking SAM sites under construction (for fear of killing Russian or Chinese advisers), from bombing Haiphong harbor (through which 85% of the DRV's military imports were delivered), and from carrying out sustained bombing that might preclude the enemy from repairing and reinforcing their air defenses. At the tactical and operational level, airpower saved countless lives but also caused "heavy civilian casualties" that may have turned the civilian population against the U.S. effort.[26] Close air support, in particular, reached a level of sophistication and responsiveness not seen since. Nevertheless, given the DRV's relatively primitive infrastructure and the support of outside powers, it remains unknown whether strategic airpower alone could have produced victory.[27]

Airpower in the Cold War

Though the "hot wars" of Korea and Vietnam dominated the headlines, the Cold War rivalry between the U.S.S.R. and the U.S. was far more dangerous. The prospect of a nuclear war that in Bernard Brodie's words would be "utterly different and immeasurably worse" held civilian and military leaders on both sides in its grip. Here, America's strategic nuclear deterrent, principally in the form of the Strategic Air Command (SAC), was by far the most important tool. For much of this period, bomber generals dominated the Air Force, which received a disproportionate share of the defense budget as well as official primacy for "strategic air

25 Young, p. 157.

26 One Air Force officer describes ". . . a climate of disregard for civilian deaths that would come to characterize airpower operations in South Vietnam." See Anthony B. Carr, "America's conditional advantage: Airpower, counterinsurgency and the theory of John Warden," *School of Advanced Air and Space Studies*, (Air University, Maxwell Air Force Base, AL, June 2009), p. 9; and Matthew A. Kocher et al., "Aerial bombing and counterinsurgency in the Vietnam War," *American Journal of Political Science, 00*, 2011, p. 5.

27 The U.S. Air Force lost 2,251 aircraft in Vietnam, including 1,737 to enemy action. 1,738 airmen were killed in action. John Schlight, *A war too long: The USAF in southeast Asia*, U.S. Air Force History and Museum Programs, 1996, pp. 103–104.

warfare."[28] Korea was dismissed as a sideshow, unlikely to be repeated and with few lessons for airmen.[29]

The strategic bomber force represented SAC's only striking arm through the 1950s, and it was mighty, disposing some 2,000 B47 medium range jet bombers and 750 B52s, as well as intermediate range Jupiter and Thor ballistic missiles that were forward deployed in Italy and Turkey. The B47 was eventually replaced in 1961 by the B58 "Hustler," in turn retired in favor of the FB111 in 1970. From 1955 onwards, SAC maintained a 24/7 posture with bombers and tankers in the air and on strip alert, both in the United States and forward deployed in Europe. From 1961–1990, SAC maintained an airborne command post, the EC135 "Looking Glass" with a general officer and battle staff, continuously in the air. At this time, it was SAC's proud boast that it served as "custodian of 90% of the firepower of the free world."[30] Inter-service rivalry was intense in this period, mostly over control and fielding of nuclear weapons and budget priorities favoring the Air Force. Political decisions to favor Air Force nuclear-armed bombers over the Navy's "super-carrier" provoked the Revolt of the Admirals in 1949. In 1955, Air Force officers claimed the Army was "unfit to guard the nation" in *The New York Times*.[31]

SAC's share of the defense budget peaked in the 1960s and declined from the 1970s on, though its importance as the first and most important line of defense remained a constant, especially after the introduction of intercontinental ballistic missiles such as the Atlas, Minuteman, and Titan. An early lead in the number of long-range missiles evaporated throughout the 1960s. From 1965–1970, the Soviets moved from a position of inferiority with 224 ICBMs to America's 934, to a clear lead with 1,440 vs. 1,054.[32]

28 Forrest E. Morgan, "The concept of airpower: Its emergence, evolution and future," in *The Chinese air force: Evolving concepts, roles and capabilities* (Washington, D.C.: National Defense University Press, 2012), p. 13.

29 Crane, p. 175.

30 Ibid. p. 180.

31 "Air force calls army unfit to guard nation," *The New York Times*, May 21, 1956, p. 1.

32 John T. Correll, "The air force and the Cold War," *Air Force Association Special Report*, September 2005, p. 13.

Despite the increasing accuracy of SAC's ICBMs and the introduction of multiple independently-targeted reentry vehicles (culminating with the deployment of the MX "Peacekeeper" in the mid-80s, which carried up to ten warheads), SAC continued to push for a replacement to the B52 in the form of the B-1. Originally cancelled by President Carter, the B-1 was revived under President Reagan. The first B-2 stealth bomber also flew for SAC, just before the end of the Cold War. With the fall of the Soviet Union, DSAC's continuous alert mission was relaxed. In 1992, SAC was disestablished. Though it never launched its missiles or bombers against the U.S.S.R., SAC represented the strongest part of the nuclear deterrent and must be credited with deterring the nuclear conflagration feared by both societies for four decades.

Airpower in the Gulf War

Operation Desert Storm, the Gulf War in 1991, presented an optimum test case for strategic airpower. The theater of operations was largely devoid of built up areas, with no forests or jungles to conceal the enemy. The Iraqi army was large but conventionally organized and relatively easy to target. Enemy air defense was not strong, and the Iraqi air force posed no real threat. The U.S. Air Force possessed a refined doctrine, and two of its principal advocates, Colonel John Warden and Lieutenant Colonel (later Lieutenant General) David Deptula, occupied key planning billets. Technology had advanced far beyond the Vietnam era, and U.S. airpower now boasted state of the art platforms.[33] First generation UAVs were now available. Air forces were enabled by GPS and secure satellite communications, as well as more advanced precision guided munitions. If ever a war could be won solely from the air, it was Desert Storm.

Warden's theories about air warfare heavily influenced the thinking of senior airmen and deserve mention here. Simplistic and easy to articulate, they promoted strategic attack as a way to paralyze the enemy

33 A partial list includes the F-15, F-16, F/A-18, F-117, B-1, B-2, the E-3 AWACS, the RC-135 Rivet Joint, and the E-8 JSTARS.

from the air by attacking key centers of gravity, leading to system failure. These were represented as five layered rings: regime leadership, organic essentials (energy, food supplies), key infrastructure (roads, rail, airfields), population, and an enemy's fielded forces. Warden theorized that destruction or negation of centers of gravity would invariably cause system collapse, with the inner ring—the regime's leadership—being the most important.[34]

In Warden's view, the use of airpower against fielded forces represented its least strategic and least effective application, for success against the inner rings would render an adversary's military forces "useless appendages." Though repackaged, Warden's thinking drew directly on the work of early airpower theorists who argued that airpower was best used strategically against "vital material targets" deep in an adversary's homeland.[35]

Critics pointed out that Warden's "hoary maxims" were overly simplistic, lacked empirical data, and were analytically inadequate, lacking any appreciation for a society's ability to adapt and react. As one Air Force officer put it, "strategists cannot allow a quantitative focus to obscure their understanding of the human interaction that constitutes both war and politics." Later, Air Force Chief of Staff Ronald Fogelman cautioned against "unfilled promises and false expectations relative to what airpower could and could not do."[36] Still, Warden remains the most influential airpower theorist of the modern era, with many acolytes in key uniformed positions today.

Warden's thinking would influence, but not dominate, the use of airpower in the Gulf War. Air operations commenced on January 17, 1991, and the "air phase" of the campaign concluded on February 23. Although the stated objective of Desert Storm was the removal of Iraqi forces from Kuwait, more than 2,250 coalition aircraft ranged widely over Iraq, striking infrastructure and command and control targets. Iraqi

34 John Warden, "The enemy as a system," *Airpower Journal*, Spring 1995.

35 See Howard Belote, "Warden and the Air Corps Tactical School," *Airpower Journal*, Fall 1999.

36 Ibid.

air defenses and the Iraqi air force were quickly overcome at low cost; only forty-two coalition aircraft were lost due to enemy action. Damage to Iraqi power and transportation systems, as well as port facilities and oil refineries, was very extensive. However, a single attack against the Al Firdos command post in Baghdad on February 13 resulted in many civilian deaths and "ended the strategic air campaign."[37]

These "strategic" targets nevertheless accounted for only 15% of the strike sorties flown in the Gulf War. 14% were directed against targets related to gaining "air control." The remainder (56%) were attacks against Iraqi ground forces.[38] In using airpower to target Iraqi ground forces in the field, the theater commander, Army General Norman Schwartzkopf, and his air commander, Air Force Lieutenant General Charles Horner, set an ambitious goal: to destroy Iraqi field forces by 50% before the start of the ground assault into Kuwait, measured by the number of armored vehicles and artillery pieces destroyed. On Day One of the ground invasion, U.S. Central Command assessed that air strikes had taken out 39% of the Iraqi army's tanks and 47% of its artillery tubes.[39] While short of the goal, this was still a tremendous achievement. Later, the use of airpower in the close air support and battlefield air interdiction roles as the ground forces maneuvered led to synergies of devastating effect.[40]

Some aspects of the air effort fell short of the ideal, notably the ability of a resilient Iraqi command and control structure, at the strategic level, to endure punishment and adapt. True "decapitation" proved elusive. A serious effort to find and kill Iraqi mobile Scud launchers,

37 Williamson Murray with Wayne W. Thompson, *Air power and the Persian Gulf* (Baltimore, MD: The Nautical and Aviaiton Publishing Co.), p. 190.

38 Thomas A. Keaney and Eliot A. Cohen, *Gulf War air power survey*, Summary Report (Washington, D.C.: USGPO, 1993), pp. 64–65.

39 Keaney, "Surveying Gulf War airpower," *Joint Force Quarterly*, Autumn 1993, p. 33.

40 Over forty-two days, coalition airpower executed 23,400 strike missions against Iraqi ground forces, of which 1,050 were close air support (CAS) missions flown by the U.S. Air Force and 750 by the USMC. Most CAS missions were flown during the ground phase, which lasted approximately 100 hours, although a few supported SOF missions prior to the start of the ground war. Leon E. Elsarelli, "From Desert Storm to 2025: Close air support in the 25th century," U.S. Air Command and Staff College, April 1998, p. 18.

an urgent priority to halt attacks intended to draw Israel into the war, was ineffective, despite devoting up to one third of the coalition air effort.[41] In the face of heavy punishment from the air, the Iraqi army did not decamp until coalition ground forces, numbering more than half a million, attacked.

The application of airpower alone could not and did not force the Iraqi army to withdraw from Kuwait. The powerful land forces arrayed in the theater were needed for that.[42] Yet the contributions of coalition airpower, if not decisive in and of themselves, could not be downplayed. They were shown to be at least as important as the land and naval capabilities employed, and, in view of the specific characteristics of the theater, which favored the employment of airpower to an unusual degree, probably more so.[43] Particularly significant was the use of airpower in the air interdiction and CAS roles; more than "strategic bombing," these missions dominated the allocation of air assets and contributed most significantly to the outcome. Strategic airlift, in-flight refueling, tactical reconnaissance and bomb damage assessment, and intelligence, surveillance, and reconnaissance from unmanned platforms as well as JSTARS all played key roles. Six U.S. Navy carrier air wings and the 3rd Marine Air Wing, cruise missile strikes from naval platforms in the Persian Gulf, as well as significant air components from coalition partners contributed heavily to the air effort. Despite the claims of many airpower advocates that the Gulf War had been won "from the air," the crushing victory in Desert Storm came about through the massive application of coordinated and integrated air, sea, and land power. Then and now, many airmen considered that the Gulf

41 "There is no indisputable proof that Scud mobile launchers . . . were destroyed by fixed wing aircraft." *Gulf War Survey*, p. 237.

42 ". . . despite the success of airpower, the introduction of ground troops was ultimately required to bring the war to a successful conclusion." Elsarelli, p. 18.

43 During the Gulf war the Air Force lost fourteen fixed wing aircraft in combat, the U.S. Navy seven, and the U.S. Marine Corps six. The Army lost five helos. Eight fixed wing aircraft were lost to non-battle causes (USAF – 4, including a B-52 and an F-111; USN – 2, and USMC – 1) with seven Army helos.

War was won through airpower alone.[44] For many others, the doctrine of victory through airpower alone remained unproven.

Airpower in Kosovo

The next opportunity to test the theory of strategic bombing came with Operation Allied Force, the NATO campaign in 1999 to drive Serbian security forces from Kosovo. Perhaps even more than Desert Storm, the Kosovo "air war" was cited as proof that modern technology had brought Douhet's theories to life. In ten weeks, from March 24 to June 11, NATO aircraft flew more than 38,000 combat missions with 1,000 coalition aircraft. Initial air operations were focused on Serbian units in Kosovo, but poor weather and limited battlefield effects quickly saw a shift in focus to "strategic" targets in Serbia proper, such as military bases, arms factories, road and rail networks, oil refineries, bridges, and command posts. Two controversial strikes in particular drew world-wide criticism: an errant attack on the Chinese embassy in Belgrade, which killed three Chinese citizens; and a deliberate attack on the Serbian national television station, killing fourteen Serbian civilians. Overall, however, civilian casualties were low.[45] Serbian air defenses were hopelessly outmatched and only two U.S. jets were lost in combat.

The debate as to whether or not Serbian President Milosevic withdrew from Kosovo due to the effects of NATO airpower alone continues to this day. Airpower advocates cite the absence of any NATO ground combat as evidence that the pure application of airpower, at least in this instance, determined the outcome. Military historian John Keegan was unambiguous in saying that the outcome proved "that a war can be won by airpower alone."[46] Other scholars point to the Kosovo Liberation Army

44 "[t]he Desert Storm air campaign . . . [made] short work in 42 days of an Iraqi military which was then the fourth-largest military force in the world." Carr, p. 9.

45 These were estimated at 500 by Human Rights Watch. See *Civilian Deaths in the NATO Air Campaign, 12* (1) February 2000.

46 The official U.S. Air Force view is unambiguous: "NATO strategic attack (i.e. air) operations coerced Yugoslav leader Slobodan Milosevic to submit to NATO demands." See U.S. Air Force Doctrine Document 3-70, *Strategic Attack*, June 12, 2007, p. 1.

in Kosovo as well as the buildup of NATO ground forces in neighboring Albania, which posed an imminent threat of intervention on the ground, as important determinates. The British commander of NATO Forces in Kosovo, Lieutenant General Sir Michael Jackson, believed that the decision of the Russian government on June 3, 1999, to urge Milosevic to capitulate was decisive.[47] U.S. General Wesley Clark, the Supreme Allied Commander Europe, felt strongly that the imminent pressure of ground invasion "in particular" forced Milosevic to back down. [48] Often lost as Kosovo recedes from memory is the fact that most Kosovars were killed or displaced *after* NATO began bombing—not before.[49]

If opinions on the role of airpower in Kosovo in 1999 still differ, what seems clear is that the case is unique. That Serbia gave in to NATO's demands after seventy-eight days of bombing, without a ground invasion, is a fact. What lessons can be drawn remain unclear. NATO deployed massive airpower with a clear technology overmatch against a small, relatively backward state unsupported by any outside power. Whether or not airpower alone could produce similar outcomes against different opponents in different circumstances remained an open question.

Airpower in Iraq and Afghanistan

Following 9/11, the invasions of Afghanistan in 2001 and Iraq in 2003 provided the next laboratories for airpower. Both campaigns saw an initial phase featuring a conventional use of airpower *en masse*, which then evolved into counterinsurgency and a much different role for airpower. Initial success in both campaigns was attributed largely to the effects of airpower, with ground forces (in far smaller numbers than in Desert Storm) playing only a secondary role. Stealth technology and a new generation of unmanned ISR platforms now augmented the

47 Andrew Gilligan, "Russia, not bombs, brought end to war in Kosovo, says Jackson," *London Daily Telegraph*, August 1, 1999.

48 General Wesley Clark, *Waging modern war*, (New York: Public Affairs, 2001), pp. 305, 425.

49 Robert A. Pape, "The true worth of air power," *Foreign Affairs*, March/April 2004.

strongest, most advanced and most sophisticated air force on the planet. The full weight of the Department of Defense was placed behind "defense transformation," a process built on concepts of network-centric warfare, sharing of intelligence, technologically-enabled situational awareness, and more precise and timely targeting. "Shock and awe," an associated concept embraced by senior defense officials, promised the rapid paralysis and defeat of adversaries through "dominant battlefield awareness," and the ability to deliver decisive force very rapidly.[50] These concepts were clearly focused on the delivery of weapons effects against "strategic" targets and not on ground operations, and they were reflected in the campaign plans for both Afghanistan and Iraq.

The U.S. response to the 9/11 attacks was swift and telling. Within weeks, special forces assisted by airpower were in the field, partnered with the Northern Alliance and battering the Taliban. On those occasions where Taliban fighters massed to confront Northern Alliance troops, U.S. airpower was devastating, often using B-52 and B-1 heavy bombers to attack tactical targets. The fall of Kandahar on December 9, 2001, and the exodus of Taliban senior leaders into Pakistan caused the collapse of the Taliban regime. The coalition effort then shifted to the Tora Bora region astride the Pakistani border, where Osama bin Laden and other al-Qaeda lieutenants were thought to be hiding. Here, dispersed and hidden targets rendered airpower much less effective. With few troops on the ground, U.S. forces were not able to prevent the withdrawal of al-Qaeda leadership to sanctuary in Pakistan.

The early stages of Operation Enduring Freedom did see one inter-service flap during Operation Anaconda, fought in the Shahi-Kot valley in Paktia province in eastern Afghanistan in January 2002. Anaconda was the first major engagement involving conventional forces, in this case a brigade task force with elements from the 10th Mountain and 101st Air Assault Divisions. In the early stages, ground forces encountered much

50 See Harlan Ullman et al., *Shock and awe: Achieving rapid dominance*, National Defense University, 1999.

fiercer resistance than expected. Lacking field artillery and heavy mortars and with only a handful of attack helicopters available, the ground force commander requested close air support. Here, the existence of separate air and ground chains of command complicated the smooth integration of CAS with ground operations, leading to controversy during and after the battle.[51]

By 2003, the Taliban and al-Qaeda had been largely driven out of Afghanistan and combat operations largely ceased. The strategic focus now shifted to the impending invasion of Iraq, which commenced on March 20, 2003. Unlike Desert Storm, which had featured several weeks of aerial bombardment before the start of ground operations, Operation Iraqi Freedom saw simultaneous air and ground operations from the outset. More than 1,700 air and missile strikes were launched on the first day. Unlike Desert Storm, when unguided conventional munitions were most often used, Iraqi Freedom saw precision guided munitions used 68% of the time in the first six weeks. Coalition air forces also flew more than 1,000 ISR missions. Air commanders determined that air supremacy had been achieved by April 6. The coalition included more than 1,800 aircraft, of which 863 were U.S. Air Force. Strategic airlift and aerial refueling again played crucial roles. More than 100,000 troops were moved to the theater of operations by air, while Air Force tankers flew in excess of 6,000 sorties to dispense more than 375 million pounds of fuel. In the first six weeks of the campaign, the coalition flew over 41,000 sorties, of which 24,000 were USAF. On May 1, after the fall of Baghdad and Mosul, "major combat operations" ceased.[52]

In this initial phase of the campaign, great efforts were made first to disable the Iraqi integrated air defense system and then to decapitate the Iraqi leadership and destroy the command and control systems linking the national leadership to its military forces, a key tenet of Warden's

51 Richard Kugler, *Operation Anaconda in Afghanistan: A case study of adaptation in battle*, Center for Technology and National Security Policy, National Defense University, 2007, p. 20.

52 Gregory Ball, *Operation Iraqi Freedom Fact Sheet*, Air Force Historical Studies Office, April 30, 2003.

"Five Rings" theory. Clear efforts were made to limit civilian casualties by using PGMs against fixed or static targets and through the use of "collateral damage estimates" made before the strikes, but civilian casualties numbered in the thousands. Dozens of strikes were employed to target Saddam Hussein and his key lieutenants without success. Virtually every security service, palace, and Baath party structure was hit, along with many "dual use" facilities, such as electrical power facilities and media sites. While Iraqi command and control was surely degraded, this sustained aerial attack did not cause system collapse. Of the fifty-five senior Iraqi leaders identified as "High Value Targets," none was successfully targeted from the air prior to declaring an end to major combat operations. According to the Human Rights Watch, "the dismal record in targeting leadership is not unique to Iraq."[53] Iraqi air defenses were destroyed with relative ease, but coalition ground forces were needed to take the ground and end the war.

While "strategic attack" once again failed of its promise, the use of airpower at the operational and tactical levels in conjunction with ground forces was stunning. To be sure, the Iraqi army was a shadow of its former self, with much of its inventory not replaced after the Gulf War. As before, weak air defense and open terrain favored the coalition.[54] The U.S. and key coalition partner Great Britain could leverage years of active experience flying Operations Northern and Southern Watch, as well as eighteen months over Afghanistan. Coalition airpower was well rehearsed and battle ready. If airpower had not proven its ability to win wars outright, there were strong grounds to argue that in conventional state-on-state conflict, the traditional paradigm had been reversed. Where before airpower had been seen as a supporting arm to the ground battle in conflicts fought on the land, Desert Storm and Iraqi Freedom

53 According to Human Rights Watch, "the dismal record in targeting leadership is not unique to Iraq. Apparently, in both Yugoslavia and Afghanistan, not one of the intended leadership targets was killed in an air strike." See "Off target: The conduct of the war and civilian casualties in Iraq," *Human Rights Watch*, 2003.

54 Fifteen USAF aircraft were shot down during the Gulf War.

suggested something new. Now the role of the land force could be seen as forcing the adversary to mass in the open—for destruction from the air.[55]

By the early fall of 2003, in both Iraq and Afghanistan, the campaigns appeared to be well won. "Stability operations" became the watchword, and rapid transition to host nation control and a quick exit were foreseen. These hopes were dashed, first in Iraq and then in Afghanistan, when virulent insurgencies emerged. Hopes for "rapid decisive victory" gave way to the Long War, a grinding, indecisive conflict of a decidedly different character. Here, the enemy rarely massed, rarely employed conventional weapons like tanks and artillery, and rarely wore uniforms. Though primitive, with no secure communications, apparent logistics or meaningful air defense, the enemy was able to fight from sanctuary and to appear even in the heart of defended coalition areas. The improvised explosive device (IED) emerged as a deadly threat. Melting into the population and with a low technological signature, the adversary—whether Taliban or al-Qaeda, Sunni or Shia—side stepped much of our technology. New methods would be needed to apply airpower to effect in what had become full-blown counterinsurgency.

Here the advent of latest-generation ISR in the form of sophisticated, and now armed, unmanned aerial vehicles promised an answer. Long endurance, armed, all-weather systems like the MQ-1 Predator and the MQ-9 Reaper came on the scene with capabilities described as "astonishing." (The RQ-4 Global Hawk, an unmanned, unarmed high altitude surveillance platform, was also fielded in Iraq and Afghanistan.) One famous four-star general opined that these systems had "fundamentally changed the nature of warfare."[56] When employed, these systems were remarkable, with long loiter times, high survivability, and precision accuracy. Controllers now gained the ability to observe "pattern of life,"

55 In three weeks, U.S. CENTCOM employed 735 strike fighters and fifty-one heavy bombers to attack 15,592 targets with 20,733 sorties. Benjamin Lambeth, *The unseen war*, U.S. Naval Institute Press, 2013, pp. 240–241.

56 Charles Dunlap, "Making revolutionary change: Airpower in COIN today," *Parameters*, Summer 2008, p. 57.

and to select the optimum time to engage the target, ideally without civilians nearby. The age of the drone now dawned.

While a major step forward, these systems were, nevertheless, far from a panacea. The first challenge was to procure them in enough numbers to matter. Combatant commanders around the globe wanted them. Air Force leaders were slow to field them at the expense of more favored manned aircraft.[57] UAV "lines" or "caps" in the theater of war were typically controlled by officers located half a world away, at Creech Air Force Base in Nevada, not downlinked directly to units that could leverage their important capabilities or integrate them fully into their operations. As a scarce resource, they were most often provided to very senior headquarters or favored special operations units. Senior leaders at times became enamored with the "soda straw" view provided by UAVs, to the neglect of more comprehensive situational awareness. Expensive to procure, these systems were also unreliable compared to manned aircraft.[58] While a clear advance, the advent of sophisticated, armed UAVs was not a game changer. Top insurgent leaders generally operated from sanctuary in Pakistan or outside Iraq. While many mid-level network leaders were targeted by UAVs, for the most part, they were quickly replaced with little degradation to their operations. With the exception of Abu Musab al-Zarqawi, leader of al Qaeda in Iraq (killed by an airstrike based on intelligence provided by the Jordanians), HVT targeting from the air in OIF and OEF was not any more successful in the counterinsurgency phase than in the conventional phases which had preceded.

The timely and effective use of airpower in counterinsurgency was also affected by inter- and intra-service politics. Airpower doctrine for decades had called for "centralized control and decentralized execution"

57 See Robert Gates, *Duty: Memoirs of a Secretary at War* (New York: Knopf, 2014), p. 130.

58 The unit cost for the RQ-4, MQ-1 and MQ-9 is $131M, $4M and $17M respectively, with program costs at $10B for Global Hawk, $2.4B for Predator, and $11.8B for Reaper. One source described Predator, Reaper, and Global Hawk as "the most accident prone aircraft in the Air Force fleet." See Brendan McGarry, "drones most accident-prone U.S. air force craft," *Bloomberg Business*, June 18, 2012.

under the command of airmen as the best way to exploit the range, speed and flexibility of air forces. In practice, this meant that air operations for both Iraq and Afghanistan were consolidated under a three-star Joint Force Air Component Commander (JFACC) who reported to the Commander, U.S. Central Command (CENTCOM) in Tampa, Florida. (Because the "preponderance" of air assets belonged to the Air Force, this was a USAF officer throughout). The JFACC performed his functions from the Combined Air Operations Center (CAOC) in Qatar, staffed primarily with Air Force officers but with representatives from each service as well as principal coalition partners.

This practice derived from Desert Storm, when the four-star combatant commander was located forward in the theater of operations. Both OIF and OEF, however, were separate theaters of war commanded by different four stars reporting to CENTCOM. Though in accordance with joint doctrine, these arrangements meant that neither joint force commander exercised actual control over air assets, which were "apportioned" by CENTCOM and "allocated" by the JFACC.[59] The JFACC/CAOC organization was optimized for large-scale campaigns where hundreds of sorties might be flown per day. But in a COIN scenario, where infrequent, low level tactical engagements were the norm, this approach proved problematic. Here, large strike packages were not needed, as the operational requirement was not to attack fixed locations or large troop concentrations. Rather, immediate close air support was most often needed, with speed and responsiveness more valued than mass.

Other factors also complicated the use of airpower in COIN scenarios. In OIF and OEF, USAF CAS required the presence of Air Force Joint Tactical Air Controllers (JTACs) on the ground. Because of the dispersed nature of COIN operations, and also because JTAC billets were often not

59 The MNF-I and ISAF four-star commanders in Iraq and Afghanistan respectively were provided with Air Component Coordination Elements (ACCE) to assist in planning; however, "tactical control of theater-wide air assets remain[ed] at the AFCENT CAOC." Lieutenant General Michael Hostage, "A seat at the table: Beyond the air component coordination element," *Air and Space Power Journal*, Winter 2010, p. 19.

filled in theater, in many cases CAS was not an option for ground commanders because no certified JTAC was present.[60] The F-15, F-16, and B-1 platforms often used to respond to requests for air support were not designed principally for the CAS mission and could not operate in close proximity to friendly ground troops; consequently, in many cases they were relegated to high speed flyovers which the enemy quickly learned to discount.

These realities caused many tactical ground commanders to conclude that true close air support was a neglected mission for the Air Force[61] Marine units, possessing their own tactical aviation under the Marine Air Ground Task Force (MAGTF) concept, were generally able to provide for themselves. Army units relied heavily on attack helicopters, which—though slower, less survivable, less well armed and with less endurance than fixed wing fighters—were very responsive and able to engage targets very close to friendlies. In highly publicized engagements in Afghanistan, such as Wanat (2008), Combat Out Post (COP) Keating (2009), and the Ganjgal Valley (2009), friendly ground forces came close to annihilation when attacked by much larger insurgent forces. Though Air Force fighters and bombers eventually responded, they were late on the scene and unable to drop ordnance or employ their gun systems close to friendly positions. The very capable AC130 gunship, ideal for providing close aerial fires where the air defense threat is low, was reserved exclusively for SOF use and flown only at night; it played no role whatever in supporting conventional forces. Field artillery and armed helicopters played the most prominent roles, along with the dogged valor of the defenders. While CAS always remained a tool, its centralization at very high levels impeded its flexibility and usefulness. One solution—to provide each theater commander with an organic JFACC, along with streamlined and decentralized

60 The officer commanding USAF JTACS in Afghanistan in the fall of 2009 reported his organization manned at 54%.

61 The author commanded a combat brigade in Iraq in 2005–2006 and served as a senior staff officer with the 82d Airborne Division in Afghanistan in 2009–2010. These observations are his own.

authorities—was seen as inhibiting "the ability to swing assets" between OIF and OEF and to "leverage the full capabilities of the CAOC."[62]

Civilian casualties (CIVCAS) also dogged the use of airpower in COIN, although technology and doctrine allowed for far more precision than in earlier conflicts. In general, civilian casualties were inflicted from the air in significantly greater numbers than from the ground.[63] In large part, this was due to the difficulty of discriminating between the local population and insurgent fighters, as well as the frequent use of civilians as shields. Standard munitions, the smallest of which was the 500 pound bomb, were also not particularly suitable for use in these conditions. A spectacular case of CIVCAS occurred on September 4, 2009, in Konduz, Afghanstan when an F-15E, responding to a request for support from German forces, attacked two fuel tankers thought to have been stolen by insurgents. More than 100 civilians were killed.[64] During the tenure of General McChrystal as ISAF Commander in 2009–2010, use of CAS was constrained by the "Tactical Directive," intended to reduce CIVCAS and improve perceptions of ISAF as a benign presence seeking to protect the population from insurgent attacks.

Summarizing the role of airpower in Iraq and Afghanistan, it is fair to say that, in the opening stages, the operational and tactical uses of airpower were decisive. As the campaigns evolved into counterinsurgency, structures and processes did not evolve to optimize the delivery of airpower in distributed and low intensity operations where close and continuous coordination with ground forces was imperative. Even in conditions where the air defense threat was very low, airpower had limited effects in most COIN scenarios and may, at times, have worked

62 Hostage, p.19.

63 For example, in 2009 the UN reported 358 civilian deaths from ISAF air strikes compared to 238 from ground operations. See United Nations Assistance Mission in Afghanistan (UNAMA), Civilian Casualty Data, 24 February 2011.

64 See Stephen Farrell and Richard Oppel, "NATO strike magnifies divide on Afghan War," *The New York Times*, September 4, 2009. Germany's Minister of Defense and Chief of Defense Staff were forced to resign in the aftermath.

against campaign objectives by alienating the population through civilian casualties.[65] An excessive concern for the safety of aircrew and airframes may have played a role in limiting the responsiveness and effectiveness of close air support. No U.S. fighter pilots were lost due to enemy fire in combat operations lasting from October 2001 to the present day.[66]

Airpower Trends Today

A survey of past wars yields some clear trends with respect to airpower. The first is that a belief in the fundamental premise—that rapid decisive victory in war can be achieved from the air, without the necessity for ground campaigns—is alive and well. Though not enshrined in joint doctrine, "Strategic Attack" remains central to air force thinking.

> Strategic attack is offensive action specifically selected to achieve national strategic objectives. These attacks seek to weaken the adversary's ability or will to engage in conflict, and may achieve strategic objectives without necessarily having to achieve operational objectives as a precondition. Strategic attack involves the systematic application of force against enemy systems and their centers of gravity, thereby producing the greatest effect for the least cost in blood and treasure. Vital systems to be affected may include leadership, critical processes, popular will and perception . . . Strategic attack provides an effective capability that may drive an early end to conflict or achieve objectives more directly or efficiently than other applications of military power.[67]

65 "Operational trends in Afghanistan showcase an increasing reliance on CAS to proxy for ground troops . . . This is driven by economy of force, but has generated a number of high-profile collateral damage incidents that threaten the entire enterprise." Carr, p. 127.

66 19 USAF, USMC, and USN fixed wing aircraft were lost in OIF and OEF, most to accident. One A-10 was shot down by an Iraqi SAM in April 2003, though the pilot survived. One F-16 crashed while conducting a low altitude strafing run near Fallujah in November 2006, probably due to pilot disorientation. 375 helos were lost in OIF and OEF through 2009, 70 to hostile fire.

67 Air Force Doctrine Document 3-70 *Strategic Attack*, with changes, November 1, 2011. In

Here can be seen, not echoes, but a reaffirmation of Douhet. In war after war, the prospect of rapid enemy collapse through strategic bombing, at low cost, and without ground fighting or casualties, has been seductive—and chimerical. The persistent belief that "winning is not about contending with fielded forces," first promulgated by the Air Corps Tactical School in the 1930s, remains alive and well.[68] This striking fact is the more remarkable because it has so often fallen short. While airpower has proven to be indispensable and increasingly predominant in state-on-state, major theater war, by itself it is inadequate. Nor has airpower been proven as a substitute for land power. In virtually all measurable cases, military campaigns conducted on the land have required a viable ground force for success. In low intensity conflict, airpower's record suffers even more, because its inherent advantages (even under conditions of low threat) are seldom realized to decisive advantage; the passion for centralization obstructs responsiveness. Even in unusual circumstances, such as the NATO conflict in Libya in 2011, overwhelming strength in the air still required a viable ground force in the form of the resistance forces to achieve battlefield (if not campaign) success.

What accounts for this enduring belief in strategic bombing in the face of so much contradictory evidence? One explanation is corporate, not strategic. A coherent doctrine that offers the prospect of single service victory, controlled by airmen, without the need for protracted land and/or naval warfare, offers more than victory. It is a compelling argument for autonomy and freedom of action, highly prized by all large organizations. To an unusual degree, the Air Force has achieved them, both in terms of setting the conditions for employment of Air Force assets and in fielding new systems. (While an important component of their parent services, naval and marine aviation complement other service warfighting

contrast, joint doctrine devotes a single sentence to strategic attack, defined as "offensive action against targets—whether military, political, economic, or other—which are selected specifically to achieve strategic objectives. See Joint Publication 3-0, Joint Operations, *The Joint Staff*, August 11, 2011, pp. iii–22.

68 Carr, p. 125.

specialties and do not consider themselves "strategic" communities in their own right.)

Air Force systems procurement reflects this independence to a marked degree. For some three decades, the unit and program costs of next generation aircraft have exploded, driving inventories down and crowding out other investment across the defense establishment. So-called fourth generation fighters (the A-10, F-15, and F-16) entered service in the 1970s at costs of $19M, $75M, and $36M respectively and are still in service today.[69] The F-117 stealth fighter, fielded in 1983 and retired in 2008, costed out at $179M. (The F/A18 Super Hornet in use by the U.S. Navy and Marine Corps is priced at $113M per copy.) In contrast, the "fifth generation" F-22 and F-35 aircraft are priced at $441M and $138M per copy, with the F-35 program described as the most expensive defense procurement program in history, due to the planned high production run.[70] The cost to taxpayers for modern strategic bombers is even more extraordinary, with the B-1 priced at $441M per copy and the B-2 stealth bomber at an incredible $3B per plane.[71]

Aircraft procurement over the past three decades has shown clear trends towards much greater cost, greater complexity, longer lead times,

69 Determining the cost of aircraft is confusing and contradictory; the differences between "unit cost," "flyaway cost," "program cost," and "life cycle cost" obscure much. The numbers cited here reflect the Program Acquisition Unit Cost (PAUC), computed by dividing the total cost of each program by the number of units to be produced. PAUC includes research and development costs, and some support costs. It is thus not an indication of actual acquisition prices. It is, however, the simplest, clearest and most accessible measure of real weapon costs that is publicly available. Figures provided by the General Accounting Office. All costs are cited in 2015 dollars.

70 See Andrea Shalal-Esa, "US sees lifetime cost of F-35 fighter at $1.45 trillion," *Reuters*, March 29, 2012. The F-35's $135M PAUC is based on a large production run of over 2,000 aircraft; if subsequently reduced, as with the B-1, B-2 and F-22 programs, the PAUC will rise dramatically. An April 2015 GAO report on the F-35 also surfaced engine reliability and other issues that "will pose significant affordability challenges." U.S. Representative Jackie Speier subsequently cited deferred testing and associated retrofit costs as "staggering… higher than the entire cost of the F-22 program."

71 The latter figure is the more remarkable in light of the B-2's operational limitations: it can fly in contested airspace only at night, is based only in the continental U.S. (necessitating extremely long flights and very limited presence in the theater of war), requires expensive climate-controlled maintenance facilities, and carries inordinate operating and maintenance costs; according to the GAO, 132 maintenance man hours per flying hour, and $171,000 per flying hour. Conceived as a Cold War weapon to penetrate Soviet defenses to attack mobile ICBM launchers, the B-2 has seen little practical use during its service lifetime.

production delays, and cost overruns. These trends may be related to the merger or disappearance of many traditional aircraft firms, reducing competition.[72] These factors in turn lead to smaller production runs, requiring legacy aircraft like the B-52, F-15 and F-16 to remain in service for decades. The net effect is a proliferation of different types of combat aircraft, fielded across many different decades, each with its own unique maintenance, training, basing and doctrinal requirements.[73]

The troubled F-35 fielding process illustrates these dynamics vividly. Billed as a multi-role stealth fighter, the F-35 is a single seat, single engine aircraft built in three variants: a standard version for the Air Force, a Navy version adapted for carrier operations, and a vertical takeoff and landing variant for the Marine Corps. Rushed into production, the F-35 program has suffered from concurrency—the need to retrofit already-fielded programs based on problems identified in subsequent field testing—as well as manufacturing quality issues, production delays, faulty cost estimates, performance issues and consistent cost overruns.[74] These deficiencies led Secretary of Defense Robert Gates to fire the two-star F-35 program manager in 2010, declaring "a culture of endless money has taken hold and must be replaced by a culture of restraint."[75]

72 A list of firms that traditionally produced aircraft for the Department of Defense includes Fairchild, McDonnell Douglas, Grumman, Hughes, Northrop, Rockwell, Vought, Curtiss, Martin, Pratt and Whitney, North American, Consolidated, Republic, Bell, Convair, Northrop, General Dynamics, Boeing, and Lockheed. All subsequently went out of business or were absorbed by Grumman, Boeing, and Lockheed. Lockheed dominates the market and produces the F-22 and F-35. Grumman has not fielded a major combat aircraft since the 1970s (it produced the A-6, F-14 and F-111). Boeing produced the B-52, KC-135 and C-17 as well as the F/A18 E/F "Super Hornet" fielded in 1999.

73 The Air Force has in service today the B-52, B-1 and B-2, as well as the A-10, F-15, F-16, F-22, F-35, C-5, C-17, C-130, KC-135, KC-10, E-3, E-8, MQ-1, MQ-9 and RQ-4, as well as a variety of trainers and executive jets.

74 "It was reported that the F-35B and F-35C models take several complex maneuvers in order to "accelerate" to their top speed of Mach 1.6, which consumed almost all of the onboard fuel." See Dave Mujumdar, "F-35B Sea Trials Aboard the USS Wasp," *The DEW Line*, August 30, 2013. The F-35 has a reported range of 1,379 miles, a service ceiling of 60,000 feet and a weapons load of 18,000 pounds (though mounting weapons externally greatly reduces its stealth characteristics). In comparison, the Vietnam era F-4 Phantom boasted a top speed of Mach 2.2 with a range of 1,615 miles, a service ceiling of 60,000 feet and a weapons load of 18,000 lbs.

75 Secretary Gates pentagon Press Conference, February 1, 2010.

Despite these problems, the Air Force has remained firmly committed to the F-35, touting it as the "premier strike aircraft through 2040 and second only to the F-22 in the air supremacy role." Though reportedly "overweight and underpowered" and a poor dogfighter, one Lockheed senior executive predicted the F-35 will be four times more effective than the F-15 and F-16 in air-to-air combat and eight times more effective than the F-16 and A-10 in air to ground combat.[76] Other defenders argue that future air engagements will take place beyond visual ranges, rendering aircraft maneuverability less important.[77] Continued funding for the programs is even more remarkable, given the compression of defense budgets under the 2011 Budget Control Act.

Gauging these competing claims is difficult without closely-guarded classified information and, more importantly, actual combat performance against a peer adversary.[78] The truth probably lies somewhere between the very optimistic projections of industry executives and the more gloomy opinions of skeptics. It is likely that the F-22 represents a major leap in capability in the air-to-air role but, due to its cost, will exist in relatively small numbers. For its part, the F-35's ability to penetrate heavily-defended airspace and carry out air-to-air or strike missions remains dependent on overcoming numerous teething problems and sustained funding for many years. Both "can fly far fewer sorties per day than smaller, more reliable fighters such as the F-16" due to higher maintenance costs—which also reduce pilot flying hours and thus proficiency.[79] A growing concern is also that new technologies, such as very high frequency radars that can "detect stealthy aircraft at long range" and infrared sensing technologies

76 George Standridge, VP for Business Development, Lockheed Martin Aeronautics Company, cited in *Space Daily*, February 22, 2006.

77 Lieutenant General (Retired) David Deptula, USAF, cited in *Breaking Defense*, July 2, 2015.

78 "Classified simulations whose assumptions are shielded from the public may indeed demonstrate the attested results, but their foundations are outside any public scrutiny, and amount to a claim that must be taken on faith." "The F-35's air-to-air capability controversy," *Defense Industry Daily*, May 30, 2013.

79 Mike Fredenberg, "What if the world's most expensive airplanes can't defeat our enemies?" *National Review*, May 15, 2015.

that can identify target aircraft at extended ranges from their heat signatures, could degrade or nullify the huge U.S. investment in stealth.[80]

Given the rapid rate of technology change and the long acquisition cycles common to aircraft procurement, a different strategy would have been to fund F-15, F-16 and F/A-18 service life extension programs and product improved aircraft (for example, by equipping them with advanced targeting and jamming pods) over time, and using the savings to invest in long-range precision-guided munitions and future unmanned platforms—in effect "skipping" a generation of manned fighters and bombers. A modified version of this approach would see a mix of high-end F-22 aircraft and less sophisticated—but still very capable—product-improved fourth generation fighter aircraft and legacy bombers. As technology improves, the advantage will lie more and more with the munition—the air-to-air or air-to-ground weapon—and not with the dogfighting characteristics of the airframe or its ability to penetrate contested airspace. Ever more accurate munitions, armed with micro-explosives that generate greater effects at less weight, and launched from standoff distances far from enemy air defenses, promise lower costs than hyper-expensive manned aircraft as well as less risk to aircrew. The advanced technology needed to field and employ truly unmanned high-performance aircraft is coming. That technology offers aircraft with greatly enhanced performance characteristics, able to sustain far greater G forces than a human pilot and with less weight (without a cockpit and its associated systems). The huge capital expenditures that have gone into F-22 and F-35 (and that are building now in support of the next generation manned strategic bomber and fighter) and the institutional commitments made to them will ensure that this leap is deferred by several decades at least.

Looking at airpower in more general terms, it is striking that alone of the world's major powers, the U.S. fields long range strike aviation in

80 See Amy Butler, "Proliferating threats open door to F-35 follow-on," *Aviation Week*, February 2, 2015; and David Shlapak, "Equipping the PLAAF," in *The Chinese air force: Evolving concepts, roles and capabilities*, p. 204.

three of its four military services. (The air arms of the U.S. Navy and U.S. Marine Corps are each larger than all but a handful of the world's air forces.) America's "separate" air forces have deep roots that extend well before the Air Force broke away from the Army and reflect a strong organizational desire to control one's own assets in combat. In the case of the Marine Corps, the indivisibility of the MAGTF and the need of the MEF commander to shape his battlespace beyond the range of organic artillery with his own resources is cited as the compelling rationale for independent control. USMC sorties, by agreement, remain with the Marine component commander; only sorties "excess to his requirements" may be apportioned by the JFACC. This arrangement implies a lack of trust on the part of Marine commanders that the Air Force can be depended upon to provide responsive and effective support. (It was in fact just this concern that spurred Army leaders following Korea to invest heavily in rotary wing attack aircraft.)[81] Similarly, employment of naval aviation rests primarily with the naval component commander, first for fleet air defense and then for "joint maritime operations," such as sea control or maritime power projection. Only sorties not needed for these requirements are made available to the JFACC.[82] These provisions essentially restrict the use of combat aircraft to the parent service.[83]

This urgent push to field hyper-expensive aircraft is the more puzzling given the overwhelming American edge in airpower.[84] In quality

81 Crane, p. 177.

82 See Joint Publication 3-30 *Command and Control of Joint Air Operations*, 10 February 2014, p. II-126

83 While the Army has a sizeable attack helicopter fleet, its aircraft operate only in close support of Army ground formations and lack the range, speed and armament of fixed wing aviation. Army aviation is therefore not normally considered in discussions of control of air power.

84 Originally envisioned as a one-for-one replacement of the F15 (some 750 units), the F-22's high cost resulted in a low production run of only 187. The F-35, intended to replace both the F-16 and F/A18, has a larger planned run of some 1,763 aircraft for the Air Force, with 600 additional B and C models for the Marine Corps and Navy. These figures depend upon continued multi-year funding which may be challenged by continued sequestration, continued technical problems (the fleet was grounded thirteen times between 2007 and 2013), and the program's "notoriously late and over-budget performance." See David Axe, "Pentagon's big budget F-35 fighter 'can't turn, can't climb, can't run,' *Reuters*, July 14, 2014. For these reasons a full production run is no more likely than with other recent programs. As the size of the program buy comes down, the unit cost will climb, probably dramatically.

and quantity, the U.S. is by far the preponderant airpower in the world, both now and for the foreseeable future. With no allies, the Russian Federation boasts only 1,275 combat aircraft, including 1,056 fighters and 219 bombers. China's numbers are higher, with a reported total of 1,923, including 1,787 fighters and 136 bombers, although again China has no allies.[85] Although both Russia and China field fourth generation fighters, many of their aircraft are older, while their pilots receive fewer flying hours than U.S. pilots. [86] Neither possesses true strategic capacity in their air forces; lacking tankers, AWACs and ISR, they are focused on conflict scenarios on their peripheries. The U.S. total for air supremacy or strike fighters is 2,889, with 155 heavy bombers. Allied air forces (NATO allies and partners in Europe, and Asia/Pacific powers like Japan, South Korea, Taiwan, Philippines, Australia and India) potentially add many hundreds more to the air order of battle.[87] In terms of airpower enablers, like tankers, transports, ISR, and AWACs, the U.S. lead increases further.[88] In overall defense spending, the U.S. defense budget of $581B dwarfs China's with $130B and Russia's with $70B.[89] In light of these facts, the argument that the U.S. lead in airpower is tenuous and at risk overstates the case by a wide margin.[90]

85 Of the Chinese fighter inventory, 672 are older J-7 and J-8 variants based on the obsolete MIG-21 and other older Soviet fighter designs. Russian and Chinese totals include naval aviation. See *The Military Balance 2015*, International Institute for Strategic Studies.

86 China has one fifth generation fighter nearing fielding, the J-20, scheduled to enter production in 2018. Russia's first 5th generation fighter, the Sukhoi PAK FA, will enter service in 2017.

87 Within NATO, the UK adds 206 fighters, Canada 77, France 313, Germany 215, Italy 224, Spain 161, Turkey 335, Belgium 59, Poland 113, the Netherlands 74, Denmark 45, Norway 57, Portugal 30, and Hungary 14, among others. Close partners Sweden and Finland add 134 and 62 respectively. In the Asia/Pacific region, among allies and partners Japan fields 353 fighter aircraft, South Korea 488, Taiwan 416, Singapore 121, and Australia 95. Potential partners include Indonesia with 43, Malaysia with 47, Vietnam with 97, and India with 943. *The Military Balance 2015*. Numbers cited for the U.S. include reserves.

88 The U.S. fields 520 tankers to Russia's 15 and China's 14; for heavy and medium transports, 709 to 190 and 65 respectively; for heavy unmanned aerial vehicles, 517 vs unknown (but very small) numbers; and for AWACs, 108 vs 22 and 18. Ibid.

89 Ibid.

90 Typical of this genre is a recent report by the Heritage Foundation that describes the Air Force as "geriatric," "obsolescent" and "in a death spiral." See Robert P. Haffa Jr., "Full spectrum air power: Building the air force America needs," Special Report #122, *The Heritage Foundation*, October 12, 2012. This paper "argues for building an Air Force to support a force capable of meeting current and

The momentum behind high priced aircraft production is spurred not only by air generals but also by heavy congressional support and by defense industry donations to political supporters. For example, despite the controversy generated by the F-22 and F-35 programs, the Air Force is committed to a next generation Long Range Strike Bomber (LRS-B), estimated to cost $600M per copy and $60B overall. Intended to replace the B-52 and B-1, the LRS-B is scheduled for fielding in the 2020s.[91] Touted as essential to overcome area denial/anti-access (AD/A2) defenses, the program demonstrates the Air Force's continued commitment to strategic attack. The obvious targets are Russia and China.

There are at least two major objections to this thinking that should concern serious strategists. The first is the notion that the U.S. should prioritize its defense spending and industrial production on penetrating the world's strongest and densest air defense networks with manned aircraft to attack high value targets (such as nuclear delivery systems and regime leadership) deep inside an adversary's homeland. In effect, this pits strength against strength, as opposed to strength against weakness, a cardinal strategic error. If successful, the survival of the adversary's regime is placed at risk, prompting the possibility of a nuclear exchange that would yield no victor. If unsuccessful, as so often historically, the massive investment is wasted. Since 1945, the U.S. has engaged in many armed conflicts—but never against the territory of a nuclear state. America and its allies should be prepared to engage in armed conflict with major powers like China and Russia when vital interests are clearly at stake. But strategic bombing campaigns against an adversary's capital and nuclear infrastructure are far from the only choices open to us.[92]

The second objection is the opportunity cost of pouring such immense resources into smaller numbers of costly aircraft at the expense

future threats to American security without regard for arbitrary fiscal guidelines and ceilings."

91 James Drew, *Flightglobal*, June 10, 2015.

92 For alternatives see the author's "America's ultimate strategy in a clash with China," with T. X. Hammes, *The National Interest*, June 10, 2014; and "Baltic fortress: NATO confronts Russia 2016," *Royal United Services Institute Journal*, Summer 2015.

of other key capabilities. The 2015 defense budget provides a striking example, with $54 billion set aside for procurement of major weapons systems. Only $1 billion of the overall total (2%) is programmed for ground systems, with the rest going to aviation and maritime programs.[93] These trends represent an unbalancing of American grand strategy that is ill-advised and dangerous. Balanced armed forces, each dominant in their own domains and able to complement each other synergistically in joint warfare, are the coins of the realm. Anything else is self-inflicted weakness. To inflict such a state of affairs on the common defense intentionally seems, at best, poor strategy indeed.

The push for expensive aircraft procurement is understandable. Production is normally spread among many sub-contractors located in many different states, a key consideration for congressional supporters. Three of the four services have very strong fixed wing aviation communities with close ties to defense industry. Movement of senior aviation flag officers to industry and of industry officials into the Department of Defense (DOD) is routine. These dynamics ensure the links between the services, the department, the Congress and defense industries, and their lobbyists are strong enough to withstand most opposition to aviation procurement. Strategic or budgetary arguments are brushed aside relatively easily because they are advanced by weaker constituencies. Even in times of budgetary uncertainty and cutbacks, the power and momentum behind airpower system acquisition is immense.

Conclusion

For eight decades at least, airpower has been a full partner in national defense, equally important with land and seapower and, under certain circumstances, even more important. For most of that history, airmen have promoted a doctrine of strategic bombing or "strategic attack" that offered the prospect of rapid victory at low cost. This promise has in the main gone

93 Matthew Cox, "Army's acquisition strategy stuck in Reagan era," *Military Times*, March 14, 2014.

unfulfilled, though it is renewed in each generation.[94] When combined with large scale land and maritime operations, air operations have played a vital part in all armed conflicts carried on by the United States.

In recent years, the cost of new aviation systems has increased exponentially. Building on themes that emphasize innovation and cutting-edge technology, the Air Force, in particular, has successfully pursued funding and political support for next generation aircraft in support of its core belief in strategic attack. While gaining and maintaining air supremacy throughout the zone of military operations is strongly supported by all military services, the fielding of hyper expensive stealth aircraft to penetrate enemy airspace and strike deeply into his homeland is far less so. Here, the historical record is controversial, the resources required are extraordinary, and consensus among the other services and the national security establishment more broadly is divided. "Strategic attack" of near-peer nuclear powers is also inherently destabilizing and dangerous. Advances in precision guided weaponry have increased the accuracy of air-delivered weapons, but the willingness of society to countenance civilian casualties far from the battlefield is low.

From a joint perspective, thorny issues still surround the effective application of airpower. Though senior flag officers are reluctant to raise inter-service concerns publicly, differing perspectives about close air support still chafe, especially with the Army (which lacks a powerful fixed wing striking arm), whose generals have modest influence over how, when, and where CAS is applied. The Air Force, for its part, would prefer to use multi-role aircraft, designed with other missions principally in mind. This lack of a dedicated CAS platform (the Air Force has pushed hard to defund and retire the A-10) fuels historic concerns that CAS remains a low priority.[95] In the depth of the battlespace, use of strike aircraft for air

94 "At regular intervals, [airmen] claim that innovations in air technology herald an entirely new age of warfare." Young, p. 156.

95 "[L]osing the freedom to apply airpower independently to decisive ends is to lose that which pilots have striven so hard to achieve for much of the history of the airplane. Thus, close air support will always be an unwanted stepchild of the Air Force." Builder, p. 137.

interdiction in conjunction with ground forces is a hallmark of modern state-on-state warfare. However, tensions persist over who should control air assets at operational depths, while the mechanisms for coordinating such efforts remain a friction point. Centralized control of air assets also remains elusive, with the Navy and Marine Corps retaining substantive control of their powerful air arms.

In 2015, the United States possesses by far the strongest and most versatile Air Force on the planet, an advantage rendered even more striking with the addition of naval and marine aviation and the air forces of close partners and allies. U.S. airpower is so dominant that in every conflict since Vietnam, losses to aircraft and aircrew have been nil, while enemy airpower has been quickly swept aside. This dominance will not erode and does not depend on continued acquisition of fewer, very expensive stealth aircraft. Continued investment in these systems, however, will starve other services and other capabilities essential for successful joint warfare in the future.

At issue is not the importance of airpower to national security. That was settled long ago. Airpower is essential to success in war and to supporting diplomacy in peace. For all its strengths, however, airpower cannot substitute for the joint force. Land and seapower are equally important and will remain so. Accordingly, a balanced approach to U.S. grand strategy that acknowledges and supports the contributions of each of the services will best serve the common defense. Within that framework, airpower and airmen will retain with confidence a hallowed place, earned over many decades of faithful service.

9

The Role of Land Power in American National Security

The United States military is dominant in air and sea power, but found itself badly stretched on the ground during simultaneous conflicts in Iraq and Afghanistan. Since the end of the Cold War, successive administrations have watered down the requirement, once enshrined in national security strategies, to fight and win in two simultaneous major conflicts. Here the author assesses U.S. land power and concludes it is not up to the task.[1]

Historically, the primary mission of the U.S. Army is *to fight and win the nation's wars on land* as part of the unified joint force. The law defines the Army's primary role as "the conduct of prompt and sustained combat incident to operations on land."[2] Decisive land combat is the Army's core competency; the Army represents the only military service capable of conducting large-scale land campaigns. Whether at home or abroad, this core mission defines the Army's reason for being and shapes its training, organization, force structure and doctrine. Since 1991, the Army has faced strong pressures to focus on "stability" operations and counter insurgency at the expense of its traditional, core missions. Today,

1 This chapter is based on two papers published previously: "American land power and the two-war construct," Association of the U.S. Army Land Warfare Paper 106, May 2015; and "The role of the army in the common defense: A 21st century perspective," Association of the U.S. Army Land Warfare Essay No. 99-4, April 1999.

2 See the 1947 National Security Act as amended, Section 205 (e).

America's reorientation on state-on-state, near peer competition finds the Army ill postured, its officer corps poorly trained in this core competency and its force structure optimized for operations at the lower end of the spectrum of conflict.[3] Shorn of much of its armor, air defense and artillery communities and with its theater logistics largely in the hands of civilian contractors, the Army today is a pale reflection of its Cold War and Desert Storm antecedents.

The Trump administration's refocus on great power competition raises once again the question of America's ability to fight and win simultaneous major conflicts, a thorny conundrum in a time of strong pressure on defense budgets.[4] Long a mainstay of American defense strategy, the "two war construct" remained an explicit commitment until the Clinton administration. Subsequently, both Democratic and Republican administrations began to parse the requirement more ambiguously as the size of the military fell. The Clinton administration saw the first iteration of the "win-hold-win" formulation. The Bush '43 national security team articulated a requirement to "maintain the capability to defeat any attempt by an enemy—whether a state or non-state actor—to impose its will on the United States, our allies, or our friends."[5] This language was further relaxed in the Obama administration, which opined "if deterrence fails at any given time, U.S. forces will be capable of defeating a regional adversary in a large-scale multi-phased campaign, and denying the objectives of—or imposing unacceptable costs on—a second aggressor in another region."[6]

This may seem merely a question of semantics, but the difference is marked. The *Quadrennial Defense Review* (QDR) language signals a

3 *The national security strategy of the United States of America*, December 2017, p. 3.

4 William J. Perry and John P. Abizaid, Co-Chairs, "Ensuring a strong U.S. defense for the future: The national defense panel review of the 2014 quadrennial defense review," (Washington, D.C.: United States Institute for Peace), July 31, 2014.

5 The White House, *The national security strategy of the United States of America*, September 2002.

6 Department of Defense, *Quadrennial Defense Review*, March 4, 2014.

projected lack of capacity to do in the second scenario what is explicitly described in the first. An ability to "fight and win two wars simultaneously," once enshrined in public strategic documents, sends an unambiguous signal that the U.S. intends to maintain military capabilities able to meet this clear requirement. For decades, this commitment underpinned our alliance structure worldwide, reassuring friends and partners and deterring potential adversaries. The qualifying language of the last several administrations signals something less.[7]

The "two war construct" is imperative for reasons that are fairly simple to explain. The United States bases its national security on an extensive network of alliances and bilateral defense arrangements. As Michael O'Hanlon has pointed out, "the United States leads a global alliance system of more than 60 partner states that collectively account for almost 80% of global GDP and more than 80% of global military spending between them."[8] This system, which provides forward basing, overflight rights, political legitimacy and additional military forces in time of conflict, is of inestimable value, but its viability as well as its deterrent effect hinges on American credibility. Our allies will be with us—if they know we will be there for them. Should the U.S. find itself committed to one major theater war (for example, on the Korean peninsula) but unable to respond decisively in another (say, in the Middle East or on NATO's borders), then that credibility is compromised not just at the point of collision but everywhere. Put another way, an America able to intervene decisively in only one region of the world at a time is no longer a global power. So constrained, the U.S. will find it difficult to play its historical role as a guarantor of a stable global system, a role whose net effect has

7 Secretary of Defense Hagel made clear in a February 24, 2014, press conference that with the current budget "this force would be capable of decisively defeating aggression in one major combat theater—as it must be—while also defending the homeland and supporting air and naval forces engaged in another theater against an adversary." Strategic documents in the Trump administration skirt the issue, using language like "retain military overmatch" and "expand the competitive space." Here the admission that the U.S. lacks the capacity to fight two land wars is explicit.

8 Michael O'Hanlon, *Budgeting for hard power: Defense and security spending under Barack Obama* (Washington, D.C.: Brookings) 2009, p. 24.

been to bring into being, largely if not entirely through America's own efforts, a rules-based international and economic order that has widely benefited much of the world.[9]

This is why prominent outside experts continue to argue "we find the two-war construct to be as powerful as ever."[10] Fighting two wars at once is never ideal. But with our security and economic wellbeing invested in a stable international order, America can't fully control what lands on its plate. The U.S. might hope to fight no more than one war at a time. If it miscalculates and cannot cope, then the extensive network of alliances and partnerships established over decades—a core and vital interest—is in grave danger. When we are able to handle to handle two wars at once, we are far less likely to have to fight any at all.

Given the preponderance of U.S. military power, shouldn't America be able to handle simultaneous major regional conflicts? After all, U.S. air, space, special operations, naval, and amphibious forces are far stronger than any that could be fielded by an adversary, and the U.S. Army, while not the largest in the world, is potent, high quality, and experienced. What then is the problem?

The answer is readily apparent in the two simultaneous wars conducted by the U.S. in the decade following 9/11. In some scenarios (such as the closing of the Straits of Hormuz in the Persian Gulf), America's preponderant air and naval power are enough to do the job. But most of the time, the challenge will be on land, and there the premise that wars can be won from the air or from the sea alone has been shown to be false.

The conflicts in Iraq and Afghanistan pitted U.S. and coalition forces against fairly primitive, low-tech opponents who could not begin to match our numbers and weapons technology. Except in the opening stages, the enemy generally refused to mass, defeating attempts to bring our

9 Christopher Layne, "Rethinking American grand strategy: Hegemony or balance of power in the 21st century," *World Policy Journal*, November 1998, p. 15.

10 "[t]he U.S. military must have the capability and capacity to deter or stop aggression in multiple theaters—not just one—even when engaged in a large-scale war." *NDP Report*, p. 2.

technology to bear decisively. Air and naval power, while enabling land forces to enter the theater of war and supporting them logistically, played only supporting roles in combat operations. They could not substitute for land power in holding, securing or dominating terrain or protecting civilian populations. Neither war could be called high intensity, yet the debilitating effects of many years of conflict proved to be extraordinary, especially for army units that deployed for year-long tours and then returned only a year later.[11] Political and military leaders could never find the ground forces needed to prosecute both campaigns effectively. As the Chairman of the Joint Chiefs famously said in 2007, "in Iraq we do what we must; in Afghanistan we do what we can."[12]

This painful experience should give pause to those who argue that other instruments—nuclear deterrence, air and seapower, special operations forces (SOF), coalition partners, or America's large National Guard formations—can offset the requirement for robust and capable active duty ground forces. These alternatives did not suffice in Iraq and Afghanistan, limited conflicts with a low air defense threat. Each has a critical role to play, but their roles are far from interchangeable. The American nuclear force is formidable but useful primarily to deter the use of nuclear weapons. Air, sea, and land power enjoy primacy in their operational domains, and can often magnify their effects when used synergistically as part of the Joint Force, but none can substitute for the other. High technology "transformative" systems can deliver better situational awareness and improved precision strike but have not proven to be decisive in their own right and carry inordinate acquisition costs. Sizeable Reserve

11 Iraq and Afghanistan severely stressed the Marine Corps as well, but Marine units typically deployed for seven months, and were generally able to retain a two-to-one deployment cycle (i.e., home for fourteen months for every seven deployed).

12 Admiral Mike Mullen, testimony before the House Armed Services Committee, December 11, 2007.

Secretary Gates echoed this view in his memoirs. "With the surge in Iraq and 160,000 troops there, the Army and Marine Corps didn't have combat capability to spare. My intent upon becoming secretary had been to give our commanders in Iraq and Afghanistan everything they needed to be successful; I realized on this initial visit to Afghanistan I couldn't deliver in both places at once." Robert M. Gates, *Duty* (New York: Alfred Knopf, 2012), p. 200.

and National Guard units were mobilized for Iraq and Afghanistan, but the political costs were high, train up periods were lengthy, and impact on civilian careers was severe.[13] Even with significant growth in Army end strength and augmentation by Marine and coalition forces, the active Army was stretched to the breaking point to execute what could be described as one medium-sized regional contingency (Iraq) and one lesser regional contingency (Afghanistan). Extraordinary steps, such as deploying soldiers multiple times for twelve and fifteen month tours, could not provide enough ground forces to achieve decisive effects in both theaters at once.[14]

The experience of more than a decade at war demonstrates, first, that simultaneous conflicts are quite possible, and, second, that dominant air and seapower is not enough.[15] America needs an army able to fight two land wars at once.[16] Ground forces in being are the ultimate currency

13 Large National Guard combat forces, up to twenty-eight combat brigades, exist in the force but require lengthy mobilization and would not be available to participate in a near-term crisis. "No large RC brigade combat teams (BCTs) or combat aviation brigades have deployed as full brigades in the first year of a global contingency in more than 50 years," the RAND Corporation, *Assessing the army's active-reserve component force mix*, October 20, 2013, p. 2. Their sustained use also raises political questions which, in all but the most serious scenarios, are problematic. For Iraq and Afghanistan, the Department of Defense adopted a "1:5" rule for National Guard brigades, meaning that after one 12-month deployment, each would be "fenced" for the next five years.

14 With 180,000 US and coalition ground troops, Multinational Forces Iraq (MNF-I) fielded a strong presence in only a handful of Iraq's eighteen provinces. In Afghanistan (a country with 215,000 more km of surface area than Iraq), 130,000 US and coalition troops could provide moderate coverage in the east and south, and only modest presence in the quieter north and west. In contrast, coalition forces in the Gulf War totaled 956,600, of which 73% or 699,000 were U.S.

15 American dominance in air and seapower is impressive. The U.S. fighter and bomber inventory in all services totals 2,643, a crushing superiority over China with 1,089 and Russia with 1,393 (both of whom would likely fight without allies). Ready reserves add another 720-combat aircraft to the U.S. total. The U.S. Navy operates ten fleet carriers, all nuclear powered; no other country has even one (the French *Charles de Gaulle*, with a displacement of 37,000 tons, is far smaller than U.S. fleet carriers with displacements above 100,000). The U.S. has fifty-eight nuclear-powered attack and cruise missile submarines—again, more than the rest of the world combined. Eighty Aegis-equipped surface combatants carry roughly 8,000 vertical—launch missile cells, outmatching the next twenty largest navies. All told, the displacement of the U.S. battle fleet exceeds the next thirteen navies combined, of which eleven are allies or partners. See Secretary of Defense Robert Gates's prepared remarks to the Navy League, National Harbor, Maryland, May 3, 2010, and *The military balance* (London: The International Institute for Strategic Studies), March 2014.

16 "Senior decision makers across five Administrations, Republican and Democrat, have been unable to avoid the reality that, in a world of continuing globalization and growing political and military uncertainty, the U.S. needs a military that is large enough and has a sufficient range of

when it comes to signaling intent and driving decisive outcomes on land. They do not do it alone but always as part of a joint force. But just as airpower is dominant in the aerospace domain, and seapower in blue water, land power dominates on land.

How much land power is needed to support this construct? Modern Army and Marine divisions pack a terrific punch, far more than in WWII, but their ability to control terrain is limited by their numbers of soldiers and the range of their direct fire systems (long range or indirect fire systems, like rocket artillery, attack helicopters, and tubed artillery, generate powerful weapons effects but cannot seize and hold ground).[17] Particularly in open terrain and with strong air support, U.S. ground forces can rupture and defeat larger enemy forces, as seen in the Gulf War in 1991 and the invasion of Iraq in 2003. But controlling terrain—"owning the ground"—is a different matter. For this reason, numbers are important.

Some idea of the scope and scale of major theater war can be seen in the size of potential adversaries. In a Korean war scenario, planners must contend with a North Korean army of 1,020,000 with 600,000 reserves and, in the event of intervention (as happened in 1950), a Chinese army of 1,600,000 soldiers with a further 520,000 reserves. In the Middle East, confrontation on the ground with Iran would engage a land force of 475,000 backed by 350,000 reserves. In Europe, Russia fields an active army of 285,000 supported by 519,000 paramilitaries and up to 2,000,000 reservists.[18] Depending on the scenario, the U.S. might fight with substantial support from allies (particularly in Korea). On the

capabilities to cover multiple major military contingencies in overlapping time frames." Daniel Goure, "The measure of a superpower: A two major regional contingency military for the 21st century," *Heritage Foundation Special Report No. 128,* January 25, 2013.

17 USMC three-star operational headquarters are considered corps headquarters equivalents and division totals include USMC divisions, which are larger than Army divisions but with fewer tanks and artillery systems (a factor largely offset by the superb close air support organic to the MAGTF). U.S. corps also include large artillery, aviation, logistical and other support units.

18 U.S. reserve forces will also come into play for extended conflicts, but not as quickly. See *The military balance*, 2014.

other hand, we must project and sustain the land force at great distances from the homeland, requiring large numbers of support troops and fewer combat troops, while likely adversaries will fight much closer to home with shorter lines of communication and a higher tooth-to-tail ratio. Our tolerance for casualties must also be far lower than our opponents, a very real constraint.

Historical examples of major theater war in the modern era similarly portray a need for robust land forces. U.S. forces in the Korean war totaled three U.S. corps with nine U.S. divisions; in Vietnam, three U.S. corps with ten U.S. divisions; in the 1973 Yom Kippur War, two Israeli corps ("fronts") with eight divisions; in the 1982 Israeli invasion of Lebanon, eight divisions under three "sector" commands; and in the Gulf War, three U.S. corps and one USMC corps-level headquarters, with nine U.S. divisions. In all U.S. conflicts, many allied units also participated.[19] Each of these single MTWs saw a crushing superiority in the air by the U.S. or Israeli air forces but nonetheless required ground forces in numbers almost as large as the entire current U.S. Army.

The U.S. Army and Marine Corps can, of course, fight outnumbered and win. Still, these numbers are sobering. At some point, force sizing begins to matter very much. Each scenario will be different, but an accurate estimate for the ground force needed to fight and win two major theater wars simultaneously (based on historical examples and the size of likely adversaries), is twelve Army and two Marine divisions of three maneuver brigades/regiments each, with an appropriate number of operational (corps level) headquarters and support units. As a "force sizing construct," this equates to seven divisions and two operational headquarters per conflict. For the Army, this means an end strength of approximately 540,000.

By way of contrast, the Army is currently sized at well below 500,000

19 Better technology does not invalidate these historical examples because the ability of land forces to "seize and hold ground"—i.e., occupy terrain—is as dependent on troop strength as on weapons technology.

and the Marine Corps at 180,000, yielding a land force of some ten Army and two and two-third Marine divisions and four deployable corps or MEF headquarters.[20] At this level of manning, most Army divisions will not be full strength. (Expanding the Army is always an option, but can't be done realistically in short time frames.) The Army also lacks a readily available pool of combat replacements, has lost much of its artillery and armor communities, and, due to the massive reductions that followed the end of the Cold War, finds itself heavily reliant on civilian contractors for much of its logistical support, a parlous liability in high intensity warfare. This force is well short of what would be required to fight two MTWs at once.[21]

The foregoing suggests that both General Powell and General Shinseki, both Army four-stars, were correct when they opined, a decade apart, that post-Cold War U.S. strategic commitments required an active Army of twelve fully manned divisions.[22] Though not a massive increase, since the drawdown of the mid-90s, this has been deemed "undoable." Why?

The answer lies principally at the nexus between DOD acquisition programs and electoral politics. Since the Vietnam War, at least, system procurement costs have soared, far outstripping the rate of inflation.

20 Department of Defense, *Quadrennial Defense Review*, March 4, 2014, p. ix. This force includes the Army's III Corps (1st Cavalry Division, 1st Armored Division, 1st Infantry Division, 4th Infantry Division) and XVIII Airborne Corps (82d Airborne Division, 101st Air Assault Division, 10th Mountain Division, 3rd Infantry Division) plus the 25th Division in Hawaii and the 2nd Infantry Division in Korea. I Corps headquarters, based in Washington state, is manned at 50% strength, with the remainder in the reserve component. The U.S. Marine Corps has two 3 star (corps equivalent) operational headquarters, I Marine Expeditionary Force (MEF) and II MEF, each with an associated Marine division, air wing, and logistics command. The USMC III Marine Expeditionary Force headquarters and III Marine Division are partially manned and would likely not participate in large scale conflicts inside of a six month or so planning horizon. The USMC has no operational headquarters above the MEF. The U.S. Army has two field army headquarters (Eighth Army in Korea and Third Army, oriented on the Middle East) which can command two or more corps.

21 Goure argues that ten Army divisions are sufficient to fight two major regional contingencies, a conclusion at odds with our experience from 1950 through 2010.

22 General Powell as Chairman of the Joint Chiefs made this "Base Force" recommendation as part of the Bush '41 administration's post-Cold War strategic review in the early 1990s, while General Shinseki, as outgoing Army Chief of Staff in 2003, declared in his final missive to the force "beware a twelve division strategy and a ten division Army."

Modern ships and aircraft are fantastically expensive and complex, consuming most of the DOD acquisition budget.[23] Defense industry has consolidated into a few giants, driving out competition and outsourcing across dozens of congressional districts to lock in political support. Unneeded bases around the country also consume budget share but cannot be closed due to congressional opposition. The Army fares poorly in this competition because its programs cost far less than jets and warships, are typically not spread across the country, and are correspondingly less attractive when bidding for congressional support.[24]

Rising personnel costs since 9/11 also placed the manpower intensive Army squarely in the budget crosshairs once Iraq and Afghanistan wound down and sequestration through the Budget Control Act hit. Manpower costs, and, in particular, health care costs, now account for approximately one third of DOD's total budget.[25] Some of the increase is related to hazardous duty, family separation, and other special pays associated with the war on terror, as well as generous family care services provided for junior enlisted service members who, a generation ago, were overwhelmingly not "married with children." Pay raises that exceeded the rate of inflation also contribute heavily. But the most serious cost driver is military health care.[26] Even so, as recently as 2012, the military personnel account in the defense budget stood at 23.3%, compared to 25.4% in 2001, and well below 1991, when it stood at 31.8%.[27] According to White House

23 "The Defense Department plans to spend more than $54 billion on major weapons systems in fiscal 2015 . . . 70 percent [will] go toward aircraft and maritime systems, from the F-35 Joint Strike Fighter and the P-8A Poseidon naval patrol plane, to the Virginia-class submarine and Aegis-class destroyer. Just $1 billion of the overall total, or about 2 percent—the lowest percentage of any category for acquisition programs—would go toward ground systems." Matthew Cox, "Army's acquisition strategy stuck in Reagan era," *Military Times*, March 14, 2014.

24 For example, for the price of a single B2 bomber, the U.S. Army could equip its entire planned armor force with new M1A2 main battle tanks. No Army program is included in the top 10 acquisition programs; the first to appear is the UH60M helicopter program which ranks fifteenth. "FY2015 weapon systems factbook," *Center for Strategic and Budgetary Assessments*, September 4, 2014, p. 2.

25 "Trends in military compensation," Bi-Partisan Policy Center and American Enterprise Institute, July 2014, p. 7.

26 Military pay has increased faster than the rate of inflation since 2000. See "Costs of military pay and benefits," *Congressional Budget Office*, November 2012, p. 3.

27 Table 6-11, DoD Outlays by Public Law Title, *National Defense Budget Estimates for FY2015,*

and DOD public documents, the overall defense budget grew more than the personnel account from 2000 to 2012.[28] The foregoing suggests that while manpower costs are a key consideration, and controlling them is imperative, the choice is far from "either/or." Disproportionate cuts in Army forces— already the weakest tool in America's strategic arsenal—to preserve expensive aviation and warship procurement programs of questionable value deserve a hard second look.

Resourcing a twelve division Army with its supporting infrastructure would mean reallocating defense dollars towards more strategically relevant priorities and away from others. (Some but not all of the manpower could be found by paring down headquarters staffs, which have ballooned in recent years.) One way to calculate cost is to divide the Army's base budget ($125B in 2014) by the number of its active divisions, yielding a figure of $12.5B per division. However, this figure overstates the case, as the portion of the Army budget devoted to supporting the Army National Guard, operating the Army Corps of Engineers with its thousands of civilian construction projects, and much of the Army's institutional base—to mention just a few—would be unaffected either way by adding two divisions. Too, Army reserve stocks of tanks, infantry carriers, tactical vehicles, helicopters, artillery howitzers and much besides are ample to equip new divisions. The actual cost of fielding an additional Army heavy division is closer to $6–8B per year.[29] This represents slightly more than 2% of DOD's current Total Obligational Authority. Given the huge sums programmed for new planes and ships, a twelve division Army is great value for the taxpayer, more strategically relevant, and well within the logic of the current defense budget.

Office of the Under Secretary of Defense (Comptroller), Department of Defense, April 2014, p. 150.

28 "DoD Emphasizing Personnel, But Overall Costs Up," *Stars and Stripes*, November 25, 2013.

29 According to senior civilian experts with the Department of the Army, the standard cost planning figure is $1.2B for every 10,000 soldiers for personnel costs and individual equipment. Planners assume that for a 15,000 soldier division added to the operating force, an additional 5,000 will be needed in echelons above division and the institutional base, yielding personnel costs of $1.8B per division.

How likely is it that the U.S. would find itself in two wars at once? There is no crystal ball, but simultaneous conflicts in the Middle East and Asia/Pacific, or the Middle East and Europe, are well within the realm of possibility. Indeed, simultaneous conflicts in Iraq and Afghanistan did materialize and, though smaller than MTW's, came close to exhausting the U.S. military. Here, deterrence is a key consideration. A strong, viable ground force yields important political dividends, reassuring allies and partners and making conflict less likely. When conflict does occur, a full strength, rapidly deployable force of twelve Army and two Marine divisions is adequate to fight and win in two separate major contingencies as the land component of the Joint Force.[30] This force is also better postured to cope with an MTW that persists beyond six months (as historically most have), providing a pool of fresh divisions that can be rotated in, as well as multiple smaller contingencies such as stability or humanitarian relief operations.[31] Should either conflict persist beyond twelve months, the nation's strategic reserve in the form of the National Guard (up to eight divisions) can be brought into play. Should requirements exceed that, the National Command Authority and the Congress may direct full mobilization and a return to the draft.

The foregoing is by no means a call for strategic primacy for land forces. Airpower must and will remain the crown jewel of America's military establishment, and as a maritime nation and the guarantor of the global commons, our seapower is essential to the common defense and our alliance structures. Yet, land power deserves an equal place. This strategic balance is all-important. Without it, deterrence falters, our options are diminished and our global reach limited. In times of peace, there is a high propensity to "accept risk" with land forces, as we see now with dramatic reductions in the size of the Army, in order to bring

30 Traditionally, and for sound reasons, the Army maintained a balance between its heavy and light divisions and should going forward as well. A predominantly light force, while cheaper, lacks the firepower, mobility and survivability to prevail in most high intensity scenarios.

31 Such as the Dominican Republic, Grenada, Somalia, Haiti, Bosnia, and Kosovo.

down defense spending while still protecting congressionally supported acquisition programs.[32]

Continued support for hyper-expensive aircraft highlights the fact that real strategy is not always at the heart of defense budgeting.[33] Even as defense spending dropped precipitously in 2012, spending on politically protected programs went forward, starving the ground force of resources and driving Army end strength to pre-WWII lows. This is not a new phenomenon, as the Army has seen every major acquisition program for the past thirty years cancelled.[34] Expanded defense spending under the Trump admiration has stabilized but not materially increased Army end strength. Its major combat systems, conceived in the 1970s and fielded in the 1980s, have no follow-ons; they will continue as the backbone of the ground force for decades to come. Where the Air Force, Navy, and Marine Corps are by far the strongest in the world, projections will yield an active Army capable of fighting one major regional contingency, with many months needed to generate enough forces to fight a second.[35]

Decades of experience in many conflicts have shown that, while air and naval forces are indispensable to success, ground forces remain necessary to seize, hold, and control the land and the populations and resources found there.[36] To overawe our adversaries, or to prevail decisively should

32 "The QDR takes risk in the capacity of each service but most notably in land forces." General Martin Dempsey, *Chairman's assessment of the Quadrennial Defense Review*, 2014 QDR, p. 61.

33 The Joint Strike Fighter program, plagued with reliability and cost overrun issues, has been described as "the most expensive program in history." Nearly $400B has been invested to date, with a planned life cycle program cost of $1.5 trillion. David Francis, *The Fiscal Times*, October 31, 2013. The Navy is currently fielding the Littoral Combat Ship, a $37B program with similar teething problems, a next-generation fleet carrier (the *USS Gerald R. Ford*) at $13B per copy, and a new class of destroyer (*USS Elmo R. Zumwalt*) at $3.5B per ship.

34 Examples include the Crusader artillery system, the Army Future Combat System, the Ground Combat Vehicle, the Grizzly combat engineer vehicle, the Armored Gun System and the Comanche armed reconnaissance/light attack helicopter. The Stryker Combat Vehicle, fielded in the early 2000s, was issued to only six of the Army's forty-five combat brigades.

35 Department of Defense, *Quadrennial Defense Review 2014* (Washington D.C.: GPO, March 4, 2014), p. ix. Army staff experts estimate that mobilizing and deploying one complete National Guard division would take up to one year, highlighting the Guard's traditional and imperative role as a strategic, but not operational, reserve.

36 A key point is that the ability of ground forces to control terrain has not materially improved with technology. The range of their systems—artillery, mortars, and the direct fire systems found in

deterrence fail, the U.S. needs a land force as strong as its air and naval brethren. A quick glance at today's strategic environment, with chaos in the Middle East, a resurgent Russia knocking on NATO's doorstep, and a rising China, show clearly that now is not the time to accept high risk. Far better to be prudent, responsible, and safe. That means a twelve-division active Army.

Today the U.S. must contend with major regional powers fielding large, well-equipped armies and theater ballistic missiles which can deliver a variety of different weapons of mass destruction. Traditional methods of deterrence, which emphasized forward presence and strategic nuclear forces, no longer apply to regional powers who face no large U.S. presence on the ground and may not fear a nuclear exchange as the possible price of breaking the peace.

In some cases, we may be able to identify these threats well in advance and structure our response accordingly. But in other cases, we may have little or no warning, as we saw in Korea in 1950 and in Saudi Arabia in 1990. While taking into account the most likely regional threats, our army must be a capabilities-based army which can respond quickly and decisively in all parts of the world, anywhere along the threat continuum. This does not mean we need an army larger and more powerful than any other. It does mean that army forces, fighting in concert with other services and allies, must be capable of meeting and overcoming those threats to our national interests that we cannot define precisely, as well as those we can. Here, staying power and striking power matter most. That means theater logistics and heavy forces.

The United States has often fought as part of an allied or multinational coalition in support of shared interests and, whenever possible,

armor and infantry units—have not advanced much since WWII. Multiple launch rocket systems and attack helicopters have far greater range, but cannot physically occupy and control terrain. The common perception that an Army division is far more capable than formerly is true in that its destructive power is far greater. But it can only control its immediate battlespace. WWI U.S. divisions, which numbered some 30,000, were far larger and, counter-intuitively, arguably more capable in this regard.

with the sanction of the United Nations. Combined operations offer political, economic, and strategic as well as military advantages which make operations as part of a larger coalition inherently desirable. Still, while the Army will often fight in concert with coalition partners, the use of American land power must not become solely dependent on outside help. Dependence on other nations runs the risk of losing that strategic independence and self-reliance which gives us the ability to control our own foreign and security policy. No nation can defend alone against *every* threat. But no nation can long endure if it cannot defend alone against *any* threat.[37]

In the next century even more than in this one, joint operations will be the standard by which we measure successful military action. U.S. national strategy demands joint doctrine and training, service interoperability, and flexible joint organizations to achieve credible deterrence and a decisive capability for victory in war and for success in military operations other than war. Historically and conceptually, joint operations dominate the strategic level of war, permeate the operational level, and influence the tactical level—but there is enduring value in service core competencies. Military power is not generic, freely substitutable, or fungible. Airpower and seapower cannot replace land power. Though modern military operations are multidimensional, encompassing land, sea, air, cyber, and space dimensions, each of the military departments has special expertise and competence in a primary dimension. No single service can win wars independently, and no service can dominate more than one warfighting dimension.

In the give and take of budget wars inside the beltway, one service or another will often press extreme claims to "strategic primacy" or exaggerate its contribution to success in past or future wars. Ultimately, however, all forms of military power are relevant only to the extent that they influence conditions and outcomes on the land. History has shown

37 In every major campaign since 1945, the U.S. has fought as part of a large coalition contributing substantial ground forces.

that air or naval power alone, while very often a necessary precondition for success on land, have never been decisive by themselves in land campaigns. To control or protect the land and the populations who live there, strong, capable Army land forces are an indispensable part of America's national military power, providing the decisive force in major joint operations conducted on the land. A clear, compelling, convincing case for the enduring importance of land power will be crucial to maintaining the strategic balance America must have to maintain its position of world leadership.

Part II

10 Presidential Decision Making and Use of Force

Case Study Grenada

The invasion of Grenada in 1983 was the first large-scale U.S. military operation since the end of the Vietnam War, and was undertaken against a backdrop of severe time pressures, the superpower rivalry with the Soviet Union, and the terrorist attack in Beirut which claimed the lives of 243 American Marines. Although a qualified success, much went wrong, providing lessons learned that informed future operations in a positive way. How and why the Reagan administration made the decision to invade this small Caribbean island shines a fascinating light on crisis decision making at the highest levels.[1]

The 1983 military intervention in Grenada was quickly followed by a flurry of articles and studies analyzing its various features and dimensions. Presidential actions under the War Powers Act, the legitimacy of the intervention under international law, media relations, foreign policy implications, and the performance of the military were all exhaustively analyzed and criticized. Yet the processes at the presidential level that ultimately led to the decision to use military force in Grenada remain largely unexplored. Though much remains unknown, enough time has passed to allow us to probe this angle with some degree of historical detachment. Operation Urgent Fury serves as an excellent case study of one administration's response to a crisis. What can the Grenada experience

1 This essay appeared in *Parameters,* May 1991.

teach us about presidential decision making and use of force at such critical times?

First, was Grenada a "crisis" at all? Some have questioned whether the events surrounding the murder of Prime Minister Maurice Bishop were only a pretext for a military intervention, already planned and rehearsed.[2] A careful review of the events leading to President Reagan's "Go" memorandum of October 23, 1983, suggests that while a predisposition to oppose further communist expansion in Latin America was undeniably present, real fears for the safety of American citizens prompted the U.S. response. Certainly, the Grenada intervention was no Cuban Missile Crisis, but the elements of perceived danger to Americans, direct Cuban and Soviet involvement,[3] and an already strained and tense superpower relationship created an explosive mixture—particularly when interpreted by an ideologically committed administration determined to avoid the imprimatur of weakness which had crippled its predecessor.[4]

Approaches to crisis management and to the broader processes of foreign and national security policymaking in the Reagan presidency were built upon the successes and failures of preceding administrations. As each incoming President does, President Reagan revised the national security process to fit his particular philosophy and operating method in light of past events.

National Security in the Reagan Administration

Ronald Reagan and his senior aides came to power determined to reduce the profile of the national security adviser (NSA) and reassert presidential control of foreign policy and national security affairs.

2 See Anthony Payne et al., *Grenada: Revolution and invasion* (London and Sydney: Croom Helm Publishers, 1984), p. 68.

3 Besides Cuban military units and advisers stationed in Grenada, the Soviets maintained a large embassy staff on the island and sponsored military and technical personnel from North Korea, Bulgaria, East Germany, and Libya. Ronald M. Riggs, "The Grenada intervention: A legal analysis," *Military Law Review, 109* (1985), p. 1.

4 Kai P. Schoenhals and Richard A. Melanson, *Revolution and intervention in Grenada* (Boulder: Westview Press, 1985), pp. 121–22.

From the beginning, the Reagan administration strove to balance commitment to "cabinet government" with centralized policy guidance and control, looking to the National Security Council staff primarily for policy facilitation and coordination.[5] The stature of the office of the national security adviser[6] was downgraded to reflect this shift in philosophy. Stripped of cabinet-level rank, the NSA was ostentatiously moved into offices in the White House basement and instructed to operate behind the scenes.[7] At least some senior NSC staffers saw their role as preserving the President's freedom to make decisions and control the policy process against attempts by other institutional actors, chiefly the State Department, to force their views on the Administration and on the policy process itself.[8]

By 1983, two national security advisers, Richard Allen and William P. Clark, had come and gone. The third, Robert C. "Bud" MacFarlane, emerged as a compromise replacement,[9] coming to the job from his position as the NSC deputy after previous service with the state department. Though outwardly committed to the cabinet government approach and to serving as mediator and honest broker for resolving conflict between State and Defense,[10] in practice MacFarlane often sided with Secretary of State George Shultz in policy disputes.[11] Clearly, he did not

5 Robert C. MacFarlane in *The presidency and national security policy*, R. Gordon Hoxie (Ed.) (New York: Center for the Study of the Presidency, Proceedings Series: Volume V, Number I, 1984), p. 260.

6 Formally referred to as the Assistant to the President for National Security Affairs.

7 Joseph G. Bock and Duncan L. Clarke, "The national security assistant and the White House staff: National security policy decisionmaking and domestic political considerations 1947–1984," *Presidential Studies Quarterly*, (Spring 1986), p. 155.

8 Constantine Menges, *Inside the national security council* (New York: Simon and Schuster, 1988), pp. 94–95.

9 Kevin V. Mulcahy, "The secretary of state and the national security adviser: Foreign policymaking in the Carter and Reagan Administrations," *Presidential Studies Quarterly*, *16* (Spring 1986), p. 296.

10 "[The President] rejected the idea that the NSC system should dominate the policy process. [He] feels that cabinet departments and agencies should play the lead role in policy development." MacFarlane, in Hoxie, p. 265.

11 Shultz's modest, self-effacing style is largely responsible for his reputation as a conciliatory, team player when in fact he emerged as the greatest single influence on foreign policy in the Reagan Administration. See Bock and Clarke, p. 165; Menges, pp. 62 and 95; and Mulcahy, p. 296.

relish an independent role, for himself or the NSC staff as a lead agency in formulating policy.

In the first Reagan term, friction between national security advisers and senior White House aides contributed to the departure of both Allen and Clark.[12] Significantly, Clark's departure occurred in mid-October 1983—leaving a newly installed MacFarlane to cope with the emerging crisis in Grenada. While the clash of bureaucratic interests undoubtedly played a role, as foreign policy perspectives occasionally collided with domestic and political considerations, concerns for personal prestige and control of access to the President were also important components of the decision setting surrounding the President.[13]

In the Reagan White House, a separate formal structure existed for crisis planning and control, reflecting the need for greater responsiveness and tighter White House control in crisis situations. While an interagency approach was retained, crisis planning groups were headed by senior White House officials. The Vice President headed the Special Situation Group providing direct, high-level policy guidance, recommendations, and options as the crisis unfolded.[14] Below the Special Situation Group (SSG), a Crisis Pre-Planning Group headed by the deputy national security adviser (in 1983, Vice Admiral John Poindexter) functioned to support the SSG with information, analysis, and supervisory assistance. For both groups, key figures routinely represented the state department, Department of Defense, Central Intelligence Agency (CIA), and NSC staff.[15]

The success of Marxist revolutionary groups in Latin America weighed on the Reagan administration heavily. Cuba was boldly providing major

12 Mulcahy, p. 291. It is interesting to note that the NSC staff was initially placed under the direct control of presidential counselor Ed Meese, principally a domestic policy adviser with a limited background in foreign, defense, or national security affairs.

13 Ibid.

14 Alexander Haig had bitterly contested the decision to have the vice president chair the Special Situation Group in 1981. His defeat reflected determination by the President and senior White House staff to retain personal control in crisis situations and contributed to Haig's loss of prestige and eventual resignation. Mulcahy, p. 289.

15 MacFarlane, in Hoxie, p. 271.

troop contingents abroad in Angola and supporting local Marxist insurgencies closer to home. The collapse of the Somoza regime in Nicaragua in July 1979 came four months after a successful Marxist coup in Grenada, led by the charismatic leader of the New JEWEL Movement, Maurice Bishop.[16] In both cases, the new regimes moved swiftly to establish close ties with Cuba[17] and the U.S.S.R., while quickly silencing political opposition.[18] A robust and dangerous Marxist insurgency in El Salvador, its prospects now improved with support from Nicaragua, complicated an already complex and worsening regional picture.

Subsequent Reagan administration actions toward Grenada cannot be fully understood without taking into account these important factors. President Reagan took office determined to reverse the pattern of American failures in foreign policy, beginning close to home in the Western Hemisphere.[19]

The Reagan administration continued and intensified the Carter "destabilization and denial" policy of economic isolation directed against Grenada, Nicaragua, and Cuba.[20] Military exercises in the Caribbean were stepped up, including the massive 1982 Ocean Venture exercise.[21] U.S. influence in the International Monetary Fund, the Caribbean Development Bank, and the European Economic Community was exerted to deny the Bishop regime external funds for development, while the state department refused to establish normal diplomatic relations with Grenada.

16 "Joint Enterprise for Welfare, Education and Liberation" party.

17 There is some evidence to suggest that Cuban special operations troops assisted in the 1979 coup led by Bishop. Dorothea Cypher in *American intervention in Grenada: The implications of Operation Urgent Fury*, Peter M. Dunn and Bruce W. Watson (Eds.), (Boulder, Colo.: Westview Press, 1985), p. 47.

18 Robert Pastor, ibid. p. 23.

19 Kai P. Schoenhals and Richard A. Melanson, *Revolution and intervention in Grenada: The New JEWEL Movement, the United States and the Caribbean* (Boulder, CO: Westview Press, 1985), p. 122.

20 Anthony Payne, *Grenada: Revolution and invasion* (London: Croom Helm Publishers, 1984), p. 61.

21 Ocean Venture 1982 involved 45,000 troops, sixty naval vessels, and 350 aircraft and was followed by the reopening of the Key West Naval Station for the purpose of enhancing command and control of U.S. forces operating in the Caribbean. Payne, p. 60.

By mid-1983, the Grenadian economy was in a shambles.[22] Mounting opposition to the socialist regime, fueled by economic collapse and the imprisonment without trial of prominent business and political leaders, began to generate severe internal pressures. In September, radical Marxist elements within the New JEWEL Movement, headed by army commander Hudson Austin and Deputy Prime Minister Bernard Coard, moved to strip Bishop of power by demanding a power-sharing arrangement leaving Coard in control of the party apparatus. In the political struggle that ensued, Bishop and a number of his supporters were seized and placed under house arrest.[23]

The Crisis Unfolds

Officials in the state department and NSC monitored the rapidly deteriorating situation in Grenada, mindful that several hundred American medical students were on the island.[24] Langhorne A. Motley, assistant secretary of state for Inter-American Affairs, later told the House Armed Services Committee that interagency groups met beginning on October 13 and that the state department had already begun coordination with the Department of Defense and the Joint Chiefs for evacuation of U.S. nationals as early as the fourteenth. On October 19, the U.S. embassy in Barbados[25] reported that Maurice Bishop and a number of supporters had been killed on orders from the Revolutionary Military Council, headed by Hudson Austin.

Interestingly, a State Department meeting was held on the morning of the nineteenth (several hours before news of the killings reached Washington, D.C.) to discuss a military rescue of the American medical

22 Schoenhals and Melanson, p. 64.

23 Ibid., p. 139.

24 Constantine Menges, special assistant to the President for Latin American Affairs at the NSC, has claimed credit for submitting a plan calling for "protection for U.S. citizens and restoration of democracy on Grenada through action by an international, legal, collective security force" as early as October 13, 1983—six days before Bishop's assassination. Clark's departure, however, delayed the submission of Menges's proposal to MacFarlane until the eighteenth. Menges, p. 64.

25 There was no U.S. embassy in Grenada.

students, a meeting held without the knowledge or participation of NSC staffers.[26] Motley testified that, from the seventeenth on, planning was conducted in an interagency forum "with representatives from all relevant agencies participating on a daily basis."[27] This is contradicted by Constantine Menges, a senior NSC staffer at the time with responsibility for Latin America. According to Menges, NSC representatives were deliberately excluded from these meetings, at least through the nineteenth, when formal crisis mechanisms came into play and formal control of events shifted to the White House.[28] Even then, a lack of enthusiasm from the Department of Defense and the Joint Chiefs,[29] combined with state department monopoly of information coming out of Grenada, gave Secretary Shultz and other senior state department officials the strongest voice in shaping the debate.[30]

Events moved quickly. By nightfall on the nineteenth, a number of significant actions had taken place, which would directly influence the ultimate decision to respond militarily. Attempts by officials in Barbados to assess the situation in Grenada in-person failed when their aircraft was turned back. Owing to the absence of any human intelligence sources on the island with a direct channel to the U.S. government, Ambassador Milan Bish in Barbados was forced to rely on sketchy information as he attempted to monitor and analyze a confused situation. State department officials were dismayed to discover that, despite the series of large-scale military exercises and an administration focus on Grenada since 1979, no specific contingency planning or intelligence preparation had been conducted for Grenada.[31]

26 Menges, p. 66.

27 U.S. Department of State, "The decision to assist Grenada," *Current Policy*, January 24, 1984, p. 2.

28 Menges, pp. 66–67.

29 Schoenhals and Melanson, p. 140.

30 Shultz spoke with President Reagan directly on, or soon after, the nineteenth to recommend a military takeover. (Ibid.) Perhaps because he was new to the job, MacFarlane deferred to Shultz and eschewed a leading role in crisis development despite the NSC's structural dominance over the crisis management process.

31 Leslie Gelb, "Shultz, with tough line, is now key voice in crisis," *The New York Times*, November 1983, Al5,p. 7

Based on Bish's judgment that events on Grenada placed U.S. citizens there in "imminent danger,"[32] an urgent meeting of the Crisis Pre-Planning Group was held on the morning of the twentieth, followed later in the afternoon by a meeting of the Special Situation Group. Officials present at this meeting, chaired by Vice President Bush, recall that a spirit of consensus prevailed in favor of a military response, given the circumstances involved. Pointed analogies to the Iran hostage crisis were drawn in this meeting, perhaps heightened by reports that Revolutionary Military Council (RMC) radicals had publicly threatened to seize American hostages several years before.[33] Even at this early stage in the crisis, discussion centered on replacing the RMC as well as securing the safety of the students. Vice President Bush directed that full-scale contingency planning proceed for a military takeover of the island, and he secured presidential approval for the diversion to Grenada of a U.S. Marine Task Force bound for Lebanon.

Much has been made of appeals for help from the Organization of Eastern Caribbean States (OECS)[34] and from Sir Paul Scoon, governor general and titular head of state for Grenada.[35] Though requests for joint military intervention in Grenada were received from the OECS on the twenty-first, state department officials in Bridgetown were in close contact with Prime Minister J.M.G.M. Adams of Barbados on the twentieth. Ambassador Bish was present at the meeting that resulted in the formal request and, along with Jamaican Prime Minister Edward Seaga,

32 Langhorne A. Motley, cited in US Congress, House, Committee on Armed Services, *Hearings on lessons learned as a result of U.S. military operations in Grenada*, 98th Cong., 2d Sess., 1984, p. 2.

33 Menges, p. 60.

34 The Organization of Eastern Caribbean States, founded in June 1981, included Antigua, St. Lucia, Dominica, Grenada, Montserrat, St. Kitts-Nevis, and St. Vincent. These members plus the much larger states of Jamaica, Trinidad, and Belize constitute the Caribbean Commonwealth.

35 Although Scoon signed a letter requesting intervention by the Organization of Eastern Caribbean States, the letter was in fact drafted in the State Department and transmitted to Scoon for signature and delivery to Prime Minister Adams of Barbados. The letter was dated October 24, but Scoon did not sign until October 26 (two days after the invasion). Scoon almost certainly did not request American intervention until prompted by State Department officials after the fact. Mark Adkin, *Urgent fury* (Lexington, MA: D. C. Heath and Co.), p. 114.

strongly urged wavering OECS representatives to move for U.S. military intervention. In fact, the formal request for U.S. help may have been drafted not in Barbados but in Washington.[36] Prime Minister Eugenia Charles of Dominica, the head of the OECS, was present in Washington on the twentieth and may also have met with state department officials.

Scoon's personal appeal to the prime minister of Barbados sought help in "stabilizing" conditions with an international peace-keeping force. It was passed to the United States on the twenty-fourth,[37] although Scoon had not asked for American military assistance or recommended a rescue mission. OECS concerns for conditions on Grenada were undoubtedly genuine, but it now seems clear that the State Department was heavily involved in framing and shaping the formal request for assistance, which emerged. A formal meeting of the National Security Council convened on the morning of the twenty-second to consider the OECS request, with President Reagan, Secretary Shultz, and National Security Adviser MacFarlane participating by secure speaker phone from Augusta, Georgia.[38] By now, contingency planning was well along, though DOD continued to express some reservations.[39] At the conclusion of the meeting, President Reagan issued a definitive decision to proceed with the military rescue and takeover of the island. A National Security Decision Directive (NSDD) to that effect was issued by mid-afternoon to the state department, DOD, and the CIA, alerting them for military operations in Grenada. H-hour was set for first light, Tuesday morning the twenty-fifth of October, subject to receipt of a final NSDD order from the President directing the operation to commence.

So far as open sources reveal, senior White House aides did not play a leading role in the process leading to the NSDD of October 22. While

36 Schoenhals and Melanson, p. 143.

37 Motley, p. 4.

38 To maintain normal appearances, the President flew to Augusta to participate in an already-scheduled golf tournament.

39 "The main impetus for a large-scale military operation came from State Department diplomats, while the Defense Department and JCS were urging delay, further study, and restraint." Michael Rubner, "The Reagan administration, the 1973 War Powers Resolution and the invasion of Grenada," *Political Science Quarterly*, *100* (Winter 1985–1986), p. 635.

they presumably conferred with the President on the subject (probably to discuss domestic political implications), as the crisis in Grenada developed, they remained background players in what was apparently a state department show. Given the President's stated desire to centralize control of crisis management in the White House, one might have expected the inner circle and the NSC staff to have dominated the decision-making process, with important input from the Pentagon and Central Intelligence Agency. In this case, the State Department retained its early initiative by feeding the Oval Office with data processed by Foreign Service personnel and embassy staff in Barbados and through Shultz's close contact with the President throughout the decision period.

At 2:27 a.m., Sunday morning, October 23, the President was awakened by MacFarlane and informed that the Marine compound near Beirut International Airport had been attacked by a terrorist car bomb and that "many" U.S. marines had been killed. Some sources cite the Beirut bombing as the catalyst that sparked the American invasion of Grenada two days later.

Although failure, or even success with high casualties, might have crippled the Reagan administration politically, the President remained determined. Reports by a CIA operative inserted on the island confirmed state department information that "no one knows who is in charge"[40] and reinforced the President in the belief that a chaotic, uncontrolled situation continued to exist which threatened the American students. While the Beirut bombing undoubtedly contributed to the crisis atmosphere and may have created an attendant desire for a countervailing political response, the evidence does not support the view that Grenada was the Administration's answer to Beirut.

Back in Washington on the evening of the twenty-third, MacFarlane briefed President Reagan on the status of planning for the operation, provided an intelligence update, and submitted a finalized NSDD for the

40 Ibid.

President's signature. Although military operations commenced some thirty hours later with the first of a number of special operations missions on the island, the presidential decision-making process culminating in the final decision to deploy combat forces ended when the President signed this document. It is interesting to note that President Reagan did not brief the attorney general or senior congressional leaders prior to making the final decision to invade Grenada.[41]

Operational control of military operations effectively passed to the Joint Chiefs at this point, although the White House closely monitored events from the White House situation room. Presidential focus now shifted to more prosaic matters, such as preparing for meetings with the media and Congress and coping with the expected fallout from the Soviet Union and other indignant Marxist and Third World states, to say nothing of offended allies.[42] The formal crisis management system continued to function, but the real decisions now lay in the fields of domestic politics and more conventional international diplomacy.

Discussion

Intervention in Grenada was not a self-contained, isolated response to a one-time, unforeseen crisis. Viewed against a backdrop of rising regional tension, strained superpower relations, and recent foreign policy failures, the ultimate decision to launch a military response should not have been the thunderclap surprise it proved to be. Yet, despite heightened scrutiny afforded Grenada following the Bishop coup and a series of military exercises in the Caribbean, both the intelligence and defense communities found themselves ill-prepared to take a leading role in shaping the developing crisis and the administration's response to it.

41 The attorney general would bear the responsibility of defending the President against charges that he failed to comply with the reporting provisions of the War Powers Act. Apparently, he was not consulted prior to transmission of the final "Go" message, nor was the congressional leadership allowed to participate in any way in the deliberations leading to the final decision. Menges, pp. 82–83.

42 Ibid., pp. 84–88.

By default, the state department, with an important agenda of its own for Latin America, moved to fill the void. Defense Secretary Caspar Weinberger and JCS Chairman General John Vessey demurred at first, not because they opposed armed intervention in principle but because of the extreme haste with which the operation was mounted.[43] It is worth noting that failure in Grenada would not have harmed the bureaucratic interests of the State Department as seriously as those of the President and the Department of Defense. This may explain, in part, the unrestrained advocacy of senior State Department officials in pushing for an immediate intervention using overwhelming force, rather than the more modest evacuation of U.S. nationals initially favored by defense officials.

Grenada demonstrated once again that congressional input into short-fused crisis decision making remains largely symbolic. As in previous crises, congressional leaders were informed only after the important decisions had already been taken. For short-term uses of force, which characterize crisis response, extra-executive actors continue to be by-standers in the decision-making process. Even within the White House itself, important figures, such as the attorney general, the press secretary, and the head of the Office of Congressional Liaison, appear not to have been consulted.

Was Grenada a "can't lose" decision, as some have argued? Clearly the elimination of the Revolutionary Military Council was a foregone conclusion once U.S. military forces were committed. Had any of a number of plausible events taken place, however, the decision to push ahead with Urgent Fury in the aftermath of the Beirut bombing could have been a political disaster of the first order. For example, had Cuban or Grenadian military forces moved rapidly to seize the American medical students, the administration could have found itself in the embarrassing position of having to withdraw in exchange for the students' safety.[44] "Horizontal

43 H. W. Brands, "Decisions on American armed intervention," *Political Science Quarterly, 102*(4), 1987, 42.

44 Several hundred students were not rescued until the afternoon of the second day, leaving plenty of time for local Marxist forces to seize them and open negotiations for their release.

escalation" in the form of Soviet or Cuban countermoves elsewhere—against Guantanamo Bay or in El Salvador, for example—could have expanded the crisis far beyond its initial compass and complicated the international situation greatly. Nor was the chance of heavy American casualties as remote as is commonly assumed. Several of the twenty-three C130 transport aircraft used in the airborne assault of Point Salines airport were holed by a 23-mm antiaircraft weapon,[45] and only a courageous decision to descend to 500 feet (the minimum jump altitude for the assaulting Rangers) averted the probable loss of several of the aircraft, each filled with over sixty troops.[46]

It is doubtful whether these contingencies were fully explored in the crisis pre-planning stages, simply because precise information about enemy military dispositions on the island was lacking. In the event, "fortune favored the brave," and major casualties were avoided. The lack of human intelligence on Grenada as the crisis developed and the lack of advanced military contingency planning placed both students and rescuers at risk, in view of the extreme secrecy and short preparation time which characterized the operation. Given the administration's focus on Grenada, the absence of this information remains something of a mystery.[47] State department advocacy of military operations appears not to have been affected by these considerations, which fall outside normal Foreign Service expertise.

During the abortive Bay of Pigs operation in 1961, the Central Intelligence Agency dominated the decision process, with the Departments of State and Defense playing secondary roles due to the CIA's monopoly of information about the operation.[48] Because of

45 Daniel P. Bolger, "Operation Urgent Fury and its critics," *Military Review*, *66* (July 1986), p. 56.

46 The decision was taken by Lieutenant Colonel Wesley Taylor, commander of the 1st Ranger Battalion, while in flight over the Point Salines drop zone.

47 Compartmentalization for operational security reasons undoubtedly degraded the decision process. Detailed intelligence on Cuban and Grenadan dispositions was available from DIA and US Caribbean Command but apparently was not used, according to a DIA representative with operational responsibilities in the area then.

48 Roger Hilsman, *To move a nation*, (New York: Doubleday, 1967), p. 31.

that operation's covert nature, other governmental agencies found themselves unable to independently assess its chances for success, thus insuring that the CIA perspective carried a stamp of authority. In the same way, the state department, in a sense, both created and sold its recommended response to the Grenada crisis by controlling and interpreting the information that reached the President, without meaningful, independent assessments by other members of the foreign policy and national security community.[49]

While the decision to intervene was both courageous and ultimately successful, its mode of implementation might well have secured a different result. The weaknesses and dangers inherent in it should have been discussed and resolved as part of the formal process. Instead, powerful bureaucratic dynamics heavily tampered with the decision-making process, relegating important formal crisis planning steps to lesser officials.

In his book on presidential decision making, Alexander George makes the trenchant observation that ready agreement between the President and his advisers on the nature of a problem and a response to it pose special dangers.[50] In the Grenada case, overreliance on a single channel of information, evaluation of a recommended course of action by a single advocate, and the absence of a senior, independent figure commissioned to rigorously explore and critique response options may all have flawed the decision-making processes that led to the decision to use force.[51]

Clearly, the President, the ultimate decision maker, participated in the few critical decisions necessary to prepare and mount the rescue

49 Senior administration officials contacted in the course of this study were (perhaps understandably) reluctant to shed light on the interagency dealings surrounding this incident. Former Secretary of Defense Weinberger referred the author to his memoirs, which contain almost no detailed account of the decision process resulting in a military response in Grenada. See *Fighting for peace: Seven years in the Pentagon* (New York: Warner Books, 1990). Former Secretary of State Shultz declined to respond to the author's request for information, citing lack of access to relevant documents. Former National Security Adviser Robert MacFarlane and former Assistant Secretary of State for Inter-American Affairs Langhorne Motley did not respond to written requests.

50 Alexander George, *Presidential decisionmaking in foreign policy* (Boulder, CO: Westview Press, 1980), p. 122.

51 Ibid, pp. 129–32.

attempt. His detachment from the details of the process, however, stands in sharp contrast to the intimate involvement of John Kennedy in the Cuban Missile Crisis or Jimmy Carter during the Iran rescue mission.[52] President Reagan's well publicized hands-off management style during the Grenada crisis preserved institutional maneuver room for the major actors involved, but it failed to provoke the crisis management system into providing the rigorous and critical debate needed to insure that all feasible options were addressed and explored.

The overriding importance of a President's personality and operating style continues to emerge in crisis after crisis. Grenada was certainly no exception. Formal crisis management groups may meet in continuous session, but once senior officials sit down in the presence of the President, normative predispositions based on background and personality and an inclination to close ranks behind the emerging consensus tend to dominate. In the Grenada crisis, President Reagan appears to have accepted each recommendation for action without critical comment.

Conclusion

What were the key ingredients of the presidential decision to use force in Grenada? The presidential personality, state department control of information and analysis, the reduced role of the NSC and its new national security adviser, and the ideological coloration of the Reagan administration were important components. The potential response of adversaries, both at the point of confrontation and elsewhere, was considered but dismissed. The perceived threat to the safety of the American medical students and the opportunity to remove a Marxist regime in the Western Hemisphere were the catalysts that impelled action.

In future crises, as in Grenada, the influences of personality and monopoly of information (and the power of interpretation conferred

52 Interview with former Director of Central Intelligence Admiral Stansfield Turner, November 23, 1989.

by it) may well continue to dominate the formal structures instituted to process and refine information for effective and informed crisis response by the President. Particularly when time is short and the pressure for action mounts, objective and rational analysis from several perspectives should inhere as an integral part of crisis response decision making. The lesson of Urgent Fury is that structure is not enough. The president, or a senior assistant chartered to do so, must provoke the system to fulfill its organizational purpose in the face of institutional opposition from competing agencies or personalities. Success in Grenada may have obscured its importance, but the lesson remains.

11 Hard Day's Night

A Retrospective on the American Intervention in Somalia

"For a great power there are no small wars."

—The Duke of Wellington

What went wrong in Somalia? This military and foreign policy disaster profoundly affected U.S. responses in the Balkans and, more tragically, during the Rwandan genocide, creating a cascading effect felt for the rest of the Clinton presidency and beyond. In fact, there were two Somalia interventions: a large, multinational humanitarian assistance mission in the waning days of President George H. W. Bush that was largely successful, and an attempt at nation building under a newly installed President Clinton that went terribly wrong. Each can yield valuable insights into complex, multinational military interventions.[1]

Almost a generation has passed since the tragic events of the October 3, 1993, when eighteen American soldiers died in the streets of Mogadishu. The fallout from Somalia was both severe and long lasting. It brought a halt to the aggressive multilateralism which initially gripped the Clinton administration, preventing any response to the Rwandan genocide which followed just months later. It limited the range of possible responses to crises in Bosnia and later Kosovo. It severely jolted the nation's confidence in its national security leadership. It shook the Clinton administration to its roots and destroyed its secretary of defense. And it induced an

1 Previously published in *Joint Force Quarterly*, 54, Spring 2009.

excessive caution and hesitancy into U.S. foreign and security policy that powerfully inhibited the administration's response to repeated acts of terrorism for the rest of the decade. In ways large and small, Somalia held American foreign policy in its grip for the rest of the decade.[2]

America lost heavily that day, in not only human terms but also international standing.[3] The causes of the disaster were both political and military and existed at every stage: at the national strategic level, where policy objectives and the goals to be pursued were fundamentally and tragically vague and ambiguous; at the operational level, where the size and composition of U.S. forces in Somalia, the command relationships established, and the missions assigned were fatally flawed; and on the ground, where secrecy, organizational rivalry, and hubris combined lethally to bring about disaster. In the years since, the heroism and fortitude of the soldiers who fought there have been celebrated. But the deeper lessons of the Somalia debacle remain painfully obscured.

Background to Intervention

American involvement in Somalia grew out of a preexisting Cold War fear of Soviet intervention in the Horn of Africa. Emerging from British and Italian colonialism in 1962, Somalia quickly succumbed to tribal strife.[4] Under General Siad Barre, military dictator from 1969 until his ouster in 1992, Somalia embraced socialism and Soviet assistance until Moscow's tilt towards Ethiopia in the mid-80's. Thereafter, Somalia inclined toward U.S. sponsorship, receiving arms and assistance before

2 The author served in Mogadishu as military assistant to Ambassador Oakley during the early phases of U.S. operations in Somalia, on detail from the White House staff.

3 It is virtually certain that America's hasty withdrawal from Somalia after the events of the 3rd of October emboldened and inspired potential adversaries around the globe. See Kenneth L. Cain, "The legacy of Black Hawk down," *The New York Times*, October 3, 2003.

4 Somali political culture is a function of the clan system; Somali clans are regionally based and trace their ancestry to a single legendary clan founder. The principal clan groupings are the Hawiye, the Darood, the Isaaq, the Rahanwein, and the Dir. Sub-clans such as the Ogadeni, the Majertain, the Habr Gidr, the Murusade and the Abgal are also highly important. To this day the major political organizations in Somalia, such as the Somali National Alliance, the Somali National Movement, the Somali Patriotic Front, and the United Somali Congress, are merely fronts for tribal or clan political movements.

degenerating into civil war in 1990.[5] In January 1991, Barre was defeated by General Mohamed Frarah Aidid, leader of the Habr Gidr sub-clan and a product of Italian and Soviet military schooling, with Barre fleeing into exile in Nigeria. A victorious Aidid occupied south Mogadishu, the capital and only major port of entry in the country. For the next year, rival clans battled for supremacy before agreeing to an uneasy ceasefire on March 3, 1992.

By that time, the international community stood horrified at the images of mass starvation beamed into its living rooms by CNN. Up to 300,000 Somalis are thought to have perished in the year preceding the cease fire. One authoritative government source reported the probable death of 25% of all Somali children.[6] In April, a small team of unarmed UN observers arrived to monitor the ceasefire, and, in August, a major UN-sponsored humanitarian assistance mission began.

Supported by U.S. flights out of Mombassa, Kenya, and a Pakistani troop presence at the port of Mogadishu, the United Nations Operation in Somalia (UNOSOM I, called Operation Provide Relief by the U.S. military) faltered quickly. Although large quantities of relief supplies arrived in Somalia, they were quickly looted or hijacked, while relief workers and nongovernmental organizations (NGOs) were assaulted and killed. Aid workers operating inside Somalia reported that food supplies were being intentionally denied to targeted populations and rival clans, spawning a man-made famine of epic proportions. In the fall, the UN reassessed its operations and called for major troop contingents from participating countries to provide military security for the humanitarian assistance mission.[7]

5 The U.S. government began to openly disassociate itself from support for Barre following the assassination of Mogadishu's catholic archbishop in July 1989 and a subsequent series of mass executions of political opponents and rival clan members.

6 This report came from Andrew Natsios, head of the United States Agency for International Development (USAID).

7 UN Security Council Resolution 794.

A Promising Start

"The people of Somalia, especially the children of Somalia, need our help. We're able to ease their suffering. We must help them live. We must give them hope. America must act."

—President George H. W. Bush

At this point President George H. W. Bush made the fateful decision to lead a large scale international intervention to halt the mass starvation which had shocked the world. President Bush seemed personally moved by the vast scale of the suffering in Somalia; as a defeated president, Bush could garner no political benefit or advantages from intervention in Somalia, and no American vital interests were engaged. His guidance was simple and direct: get in fast and stop the dying. The administration policy focused almost exclusively on providing security for humanitarian assistance, with no mention of nation building or long-term stability operations.[8]

Beginning in early December, large numbers of U.S. troops began moving towards the Horn of Africa. At month's end, more than 28,000 Marines and soldiers from the 1st Marine Expeditionary Force ("I MEF") and 10th Mountain Division had arrived.[9] A combined joint task force called UNITAF (for "Unified Task Force") was established under I MEF's commander, Lieutenant General Robert Johnston, to control all U.S. and UN forces.[10]

Based in Mogadishu, but with major elements in outlying cities like Bale Dogle, Baidoa, Oddur, Merca, and Kismayu, and supported by 10,000 coalition soldiers from twenty-four countries, UNITAF quickly

8 Robert Oakley, "An envoy's perspective," *Joint Force Quarterly*, Autumn 1993, p. 45.

9 The decision to send a corps-level MEF HQ was clearly correct, as a division would have experienced severe span of control problems with the many coalition brigades and battalions in Somalia.

10 The UNITAF staff was built around the I MEF battlestaff with significant augmentation from other services and national contingents. The Operations Officer was Brigadier General Anthony Zinni, USMC, later Commander-in-Chief of Central Command.

established order.[11] The force that went into Somalia that December was muscular and well-armed, with liberal rules of engagement that allowed U.S. soldiers to engage any armed Somalis thought to pose a threat.[12]

In addition to overwhelming military force, the American-led intervention featured a small but experienced diplomatic effort, headed by U.S. Special Envoy Robert Oakley. With former experience as a senior NSC staffer and as ambassador to Pakistan, Zaire, and Somalia, Oakley was well known to the major faction leaders and well versed in internal Somali politics and rivalries.[13] Significantly, Oakley's "U.S. Liaison Office" or USLO (in the absence of a functioning central government, there was no U.S. embassy), was sited near Aidid's personal residence in south Mogadishu and was guarded by only six U.S. Marines.[14]

On the ground, both Johnston and Oakley worked to coordinate political and military efforts to rush humanitarian assistance to threatened areas. Military officers were seconded to Oakley's staff,[15] and UNITAF provided senior, experienced liaison officers to meet regularly with USLO and with UN and NGO agencies.[16] Both military and civilian representatives worked together in Civil-Military Operations Centers in the capital and in outlying areas to plan and execute humanitarian assistance operations. Somalia was organized into large Humanitarian Relief Sectors, each placed under a capable coalition unit, to ease coordination and command

11 Many of the participating national contingents were marginally useful at best, but some—notably the Italians, French, Canadians, Australians, Botswanans, Belgians, and Moroccans—performed successfully.

12 See General Joseph Hoar, USMC, "A CINC's perspective," *Joint Force Quarterly*, Autumn 1993, p. 59.

13 At the time, Oakley was in retirement and held the distinction of being one of only four "Career Ambassadors," the highest rank in the foreign officer hierarchy.

14 In 1993 Mogadishu was effectively partitioned into the south, controlled by Aidid's Habr Gidr sub-clan, and the north, the turf of factional leader Ali Mahdi. The city was separated by the so-called "Green Line."

15 Colonel (later Major General) Rich Mentemeyer, USAF, was seconded from the CENTCOM staff to be Ambassador Oakley's Military Adviser.

16 The high quality and seniority of the Marine officers detailed as liaison officers to U. S. diplomats and UN/NGO agencies was remarkable. Most were full colonels with extensive command experience. The Marine officer detailed to serve as personal liaison between Johnston and Oakley was Colonel Mike Hagee, later the Commandant of the Marine Corps.

and control challenges.[17] Somali leaders were brought together frequently to hammer out solutions to local conflicts in meetings brokered by Oakley in the neutral setting of the USLO compound.

At the outset, Somali factional leaders were told politely but firmly that, while the intention was not to impose any particular ruler or system of government in Somalia, no armed threat would be permitted to challenge U.S. or UN troops. All "technicals"—civilian trucks and vehicles modified to mount heavy weapons—were required to be stored in monitored cantonment areas, and no weapons could be carried visibly in public.[18]

The results were immediate and dramatic. Within a month, massive amounts of food aid were flowing freely, and the death toll from starvation had dropped exponentially.[19] Armed clashes between warring factions had declined precipitously, and U.S. casualties were low. Although nominally a UN operation, Operation Restore Hope was clearly a U.S.-led effort.[20] Both Aidid and Ali Mahdi, anxious to position themselves as future national leaders with U.S. backing, generally cooperated with U.S.-sponsored initiatives to encourage local and regional cooperation.[21] In Mogadishu and elsewhere, joint councils actually emerged to manage port operations, police functions and other forms of public administration.

17 Jonathan T. Dworken, "Restore Hope: Coordinating relief operations," *Joint Force Quarterly*, Summer 1995, p. 15.

18 The term "technicals" derived from the practice by aid organizations of listing locally hired Somali security, often mounted in civilian vehicles modified to carry heavy weapons, as "technical support." These were, of course, the same Somalis who often robbed aid shipments and threatened aid workers when off duty. Numbers of technicals were moved outside to Mogadishu to avoid monitoring, but as this removed them from proximity to US and allied troops, no action was taken. A short-lived attempt was made to confiscate weapons from personal homes in early January but was quickly abandoned.

19 "On average, the military escorted seventy convoys carrying 9,000 metric tons of supplies inland from Mogadishu each month." Dworken, p. 16.

20 While the UN High Commissioner for Refugees (UNHCR) played key roles in coordinating NGO activities in conjunction with the U.S. military, the Special Representative of the UN Secretary General, Ambassador Ismat Kittani, played at best a marginal role. UN leadership was hampered by the fact that the Secretary General, Boutros Boutrous-Ghali, and Aidid disliked and distrusted each other because of previous dealings dating from Ghali's service as Egyptian Deputy Foreign Minister for the Upper Nile. In that period, Ghali had enjoyed close relations with Siad Barre.

21 Although both members of the Hawiye clan, Aidid and Ali Mahdi headed the rival Habr Gidr and Abgal sub-clans respectively and were mortal enemies. The pastoral Abgal were native to Mogadishu, the nomadic Habr Gidr were outsiders from Central Somalia who occupied south

The process was not smooth. Simmering clan tensions and occasional clashes persisted, and attempts to encourage cooperation between rival factions failed as often as they succeeded. Still, the primary task of "stopping the dying" was a major success. Throughout, the U.S. approach was consistent and focused: don't take sides, focus on the humanitarian mission, and avoid direct confrontation where possible—and when not, act forcefully and directly.[22] By the end of President Bush's term of office on January 20, 1993, death by starvation had largely ceased and open clan warfare had diminished drastically.

Change of Mission

> *"With this resolution, we will embark on an unprecedented enterprise aimed at nothing less than the restoration of an entire country as a proud, functioning and viable member of the community of nations."*
>
> —Madeline Albright

The U.S. mission to Somalia, Operation Restore Hope, changed dramatically after President Clinton's inauguration. Restore Hope was characterized by a short-term focus, overwhelming force, close cooperation and liaison between its political and military components, clear political guidance and a distinct policy of noninterference in the murky waters of local Somali politics. While attempts were made to support local and national reconciliation to ease clan rivalry and support humanitarian assistance, nation building was never allowed to emerge as a primary goal. In sharp contrast, the UNISOM II effort (dubbed Operation Continue Hope by U.S. military planners) envisioned indefinite time horizons, far weaker military forces, more ambitious and ambiguous political goals

Mogadishu during the civil war. During Operation Restore Hope, Oakley's USLO actually rented its vehicles and drivers from Aidid—prompting complaints of favoritism from Ali Mahdi.

22 On January 7, after repeated complaints about U.S. troops being fired on in south Mogadishu, UNITAF attacked one of Aidid's cantonment sites, destroying dozens of technicals and killing more than thirty Somali militia men.

and a more idealistic and ideological tone and character. Under Bush, the mission was humanitarian assistance. Under Clinton, the mission would become far more expansive.

The nature of the U.S. mission in Somalia began to change almost from the day Clinton took office. His national security team lacked experience but not confidence,[23] and within weeks of the inauguration, a strong shift in policy began to emerge. The focus now changed from "stopping the dying" to rebuilding Somali national institutions, infrastructure, and political consciousness; from the U.S. to the UN; and from overwhelming military force to the smallest possible American military footprint.[24] On the March 26, 1993, U.S. Ambassador to the United Nations Madeleine Albright voted in favor of UN Security Council Resolution (UNSCR) 814, creating a successor UN organization in Somalia, UNISOM II. Among other things, UNSCR 814 committed the UN to more expansive national reconstruction and political reconciliation goals and charged UNISOM II to disarm the Somali clans, a fateful step that presaged the failures that would soon follow.[25]

To ensure U.S. control, retired Admiral Jonathan Howe was named to head UNISOM II as the secretary general's special representative. Howe had recently served as deputy national security adviser and was, therefore, experienced in the interagency process and, presumably, read in on the complexities of the mission in Somalia. Polished and articulate,

23 The inexperience in senior government of key players in framing policy towards Somalia is key to understanding what transpired. Secretary of Defense Les Aspin and USUN Ambassador Madeleine Albright had never held senior executive branch positions. National Security Advisor Anthony Lake had served for many years in academe and took little interest in Somalia until the crisis erupted.

24 Richard J. Norton, "Somalia II," in *Case Studies in Policy Making and Implementation*, 6th edition, *Naval War College* (Newport, RI: 2002), p. 168. See also Dorcas Eva Mccoy, "American post-Cold War images and foreign policy preferences toward "dependent" states: A case study of Somalia," *World Affairs*, Summer 2000, p. 2.

25 Major Somali clan leaders and political groups met in Addis Ababa on March 27, 1993, to sign a reconciliation pact that, among other things, called for "substantial" arms reductions to begin within ninty days and a peaceful transition to a coalition government within two years. Lack of progress in turning in weapons in accordance with the Addis Ababa accords was cited as a catalyst in the aggressive UNISOM II approach, which ultimately wrecked the policy of non-alignment with Somali factions.

as a military officer he represented both non-partisanship and a willingness to take direction and follow orders.[26] Major General Thomas Montgomery, a tank officer serving on the Army staff, was named as Commander of U.S. Forces in Somalia and Deputy Commander of UNISOM II's military forces (under Turkish Lieutenant General Cevik Bir).[27] Significantly, however, UNISOM II lacked a trained military staff and important communications and intelligence systems. Even Montgomery's own U.S. combat forces were placed under CENTCOM's operational control, 7,000 miles away. An ad hoc organization beset with conflicting national agendas and interests, UNISOM II was poorly suited to conduct major combat operations. Very quickly, things began to go wrong.

Driven by a strong desire to pull U.S. forces out, the U.S. troop presence in Somalia declined from 17,000 in mid-March to 4,500 in early June as UNITAF disbanded and I MEF went home.[28] Although many coalition units remained, most of the credible combat capability resident in Somalia left with the Americans. This dramatic reduction in U.S. military force coincided with aggressive actions to force various Somali militias to disarm. As Aidid ruled south Mogadishu with his "Somali National Alliance," or SNA, where UN forces were concentrated, UNISOM II pressed the Habr Gidr hard. Predictably, there was resistance, and UNISOM II began to take casualties. Almost immediately, national contingents began to suspend

26 Howe was a career naval officer with unusually impressive politico-military credentials. He lacked, however, experience as a diplomat, particular knowledge of Africa or a background in managing complex land operations involving multiple coalition actors, agencies and interests. He was essentially a sailor with sterling inside-the-beltway, interagency credentials.

27 An outstanding officer, Montgomery came from a Cold War, Central Europe background and lacked expeditionary experience in complex third world settings. (The Army possessed a number of general officers with these credentials, notably Major General Steve Arnold, previous Commanding General of the 10th Mountain Division, who had served in Somalia with UNITAF.) Assigned to the Office of the Army Chief of Staff, Montgomery was selected primarily due to his immediate availability and prior history with the Army leadership. His dual status was an attempt to solve the problem of UN command over US troops, a traditional "redline." Any UN order to use US forces, which provided UNISOM II's Quick Reaction Force or QRF, would go through Montgomery, who would then transmit the order and command the mission as COMUSFORSOM.

28 Of these 4,500, only 600 were ground combat troops. The rest were headquarters, logistics and aviation soldiers.

activities that placed them at risk of reprisal. Increasingly, Howe and Montgomery turned to the lone remaining U.S. light infantry battalion for the hard missions.

On the June 5, in an attempt to search one of Aidid's heavy weapons storage areas, a Pakistani unit was badly mauled. In a lengthy firefight, Aidid's militia killed twenty-three and wounded fifty-nine. UNISOM II's Malaysian armor and American QRF were unable to intervene in time to prevent the heavy loss of life.[29] From that date, everything changed in Somalia.

Both the UN and the U.S. government reacted heatedly. On June 6, the UN Security Council approved a resolution explicitly calling for the "arrest and detention for prosecution, trial, and punishment" of the perpetrators of the attack on the Pakistanis.[30] Despite later attempts to distance the Clinton administration from this action, there is little doubt that the U.S. government not only supported but also forcefully promoted this response.[31]

Howe immediately requested special operations forces, and, while the Administration pondered a response, UNISOM II stepped up its operations against Aidid. In mid-June, U.S. forces attacked a radio station and ammunition dumps and attacked targets throughout the city with AC130 Spectre gunships.[32] On July 12, U.S. forces conducted a major raid on the "Abdi House," scene of a meeting of SNA leaders to discuss UN reconciliation proposals. Many were not in agreement with Aidid and were supportive of efforts to end the tribal in-fighting and encourage foreign aid and investment. Nevertheless, ground troops and Cobra helicopters

29 Prior to his departure for Mogadishu, Howe was warned in background briefings that "Aidid will test you early and hard. His most likely target will be the Pakistanis". Author personal narrative.

30 This was UN Security Council Resolution 837.

31 Norton, 178.

32 Major J. Marcus Hicks, USAF, "Fire in the city: Airpower in urban, smaller-scale contingencies," *School of Advanced Airpower Studies*, Air University, Maxwell Air Force Base, Montgomery, Alabama, June 1999, p. 74. After the loss of one Spectre gunship, caused when a 105mm round went off in the breach of the gun while in flight, the AC130s were withdrawn. A subsequent request from MG Garrison for their return was denied by Secretary Aspin.

firing heavy antiarmor missiles destroyed the building with heavy loss of life. Fifty-four Somalis were killed, and, in the ensuing rioting, four Western journalists attempting to cover the event were torn apart by the enraged crowd.[33] The Abdi house raid went far to unify Aidid's people solidly against the Americans and raised the conflict to a new level. Its importance in changing Somali attitudes is hard to overstate.

On August 8, a remotely detonated antitank mine (similar to the improvised explosive devices or IEDs commonly used today in Iraq) killed four Americans, and similar attacks on August 19 and 22 wounded ten more. Mogadishu was fast becoming a free fire zone, and, as hostilities escalated, President Clinton approved the dispatch of the Joint Special Operations Command (JSOC) along with a 440 soldier Joint Special Operations TF (JSOTF).[34]

Called Task Force (TF) Ranger, this composite unit was built around a rifle company and battalion headquarters element from the 75th Ranger Regiment, a detachment from the Army's famed Delta Force, and an aviation element from the 160th Special Operations Aviation Regiment (SOAR), equipped with MH60 Black Hawk utility helicopters and MH6 and AH6 "Little Bird" light helicopters.[35] Small numbers of communicators, Air Force combat controllers and para rescue airmen ("PJs"), and SEALs were included. TF Ranger, led by JSOC Commander Major General William F. Garrison, did not report to General Montgomery as Commander of U.S. Forces in Somalia. Instead, as a "strategic asset," it reported directly to Central Command in Tampa, Florida.[36]

Upon arrival, TF Ranger immediately went to work, conducting their first raid against "leadership targets" on August 30. Five other

33 George Monbiot, *The Guardian*, January 29, 2002.

34 Secretary of Defense Les Aspin signed the order deploying TF Ranger on August 22, 1993.

35 MH60s are standard Black Hawk utility lift helicopters modified for special operations with inflight refueling and other non-standard systems. The MH6 is a modified OH6 light observation helicopter with more power, adapted for carrying four special operations soldiers externally on sled seats mounted above the skids. The AH6 is a variant equipped with rockets and miniguns for light attack missions in support of special operations.

36 Norton, p. 172.

raids took place in September. All were based on short-fused intelligence and followed a similar tactical pattern: an insertion by MH60 and MH6 helicopters, with Rangers forming an outer perimeter and Delta operators conducting the actual prisoner snatch, supported by a ground convoy to extract detainees and covered by AH6s aloft. These operations met with mixed success. In one, Aidid's financier and right-hand man, Osman Otto, was captured. But others betrayed the spotty human intelligence available to the Americans. In separate instances, the Rangers moved against the headquarters of the UN development program and the offices of Medicins sans Frontiers and World Concern, leading aid agencies working in Mogadishu.[37] Another raid netted the former Mogadishu police chief, well known as a neutral player and not aligned with Aidid.[38] As with the raid on the Abdi house, poor human intelligence and a lack of situational awareness plagued TF Ranger operations. Significantly, there was little or no coordination between Garrison and Montgomery.

Supporting intelligence structures also deserve comment. A CENTCOM intelligence assessment team traveled to Mogadishu in June 1993 and reported that the capture of Aidid was "viable and feasible," though, in private, team members described the task as "extremely ugly . . . with numerous potential points of failure." Regrettably, the CENTCOM Intelligence Support Element (CISE) in Mogadishu experienced 100% turnover in the third week of September 1993. New arrivals were provided an "uneven" transition. JSOC intelligence officers later reported that CISE support to TF Ranger was "minimal," with a poor focus on critical human intelligence.[39]

37 Monbiot,p. 2.

38 Norton, p. 172.

39 See Colonel James T. Faust, USA (Ret), former JSOC J2 Operations Chief, in Task Force Ranger in Somalia (Carlisle, PA: U.S. Army War College Personal Experience Monograph, March 1999), p. 13.

The Gloves Come Off

> *"We have a war going on in Somalia. From a tactical and operational perspective it is not going well . . ."*
>
> —Major General Dave Mead
> Commanding General, 10th Mountain Division
> September 15, 1993

In mid-September, the Commanding General of the 10th Mountain Division, Major General Dave Mead, sent an explosive personal message to the Chief of Staff of the Army. (This message, a "P4" in military parlance, has never before been made available to the public.) Visiting his troops in Mogadishu (the UNISOM II Quick Reaction Force consisted of a helicopter task force and infantry battalion from the Fort Drum-based 10th Mountain Division), Mead was shocked at what he found.

> Mogadishu is not under our control. Somalia is full of danger. The momentum and boldness of Aidid are the prime concern. The trend lines are in the wrong direction. Thus the mission overall and the security of the U.S. Force are threatened.[40]

Mead went on to describe how hundreds of armed Somalis had attacked U.S. combat engineers and Pakistani tank crews in a major fight along the Marehan Road in Mogadishu on September 9. In that engagement, two rifle companies from the QRF infantry battalion rushed to the scene, only to be forced back to their compound under heavy fire. Despite severe losses, Aidid's militia men fought hard and aggressively that day in the face of helicopter gunships, UN armor, and several hundred U.S. infantrymen.[41]

40 Personal Message from Commanding General 10th Mountain Division to the Army Chief of Staff, 15 September 1993, p. 1. Hereafter cited as "Meade P4."

41 The American QRF fell back under such heavy pressure that rather than move to the main

As Mead grasped after only a few days on the ground, conditions in Mogadishu had deteriorated dramatically. Aidid was well aware of the American manhunt and the reward offered for his capture. On multiple occasions, he had demonstrated a readiness to take the Americans on directly, despite their advantages in firepower. The national contingents showed no stomach for the campaign to "get" Aidid—a number had in fact negotiated private agreements after the Pakistani massacre.[42] With a very limited U.S. force on the ground, UNISOM II and its American backers were in real trouble.

> This war is the United States versus Aidid. We are getting no significant support from any UN country. The war is not going well now and there is no evidence we will win in the end. We have regressed to old ways. Our efforts are not characterized by the use of overwhelming force, not characterized by a commitment to decisive results and victory, not designed to seize the initiative, and there is no simultaneous application of combat power, and not a plan to win quick. All this has the smell and feel of Vietnam, Waco and Lebanon . . .[43]

General Montgomery, the on-scene commander, apparently did not express the same level of alarm in his reports to General Hoar at CENTCOM or to UN headquarters in New York. But he was worried enough to request a major addition to his force, in the form of an American mechanized infantry battalion task force equipped with main battle tanks and artillery. This request reached CENTCOM in mid-month and was refused on the grounds that increasing the U.S. "footprint" in Somalia

entrance of the U.S. compound they blew a hole in the back wall with explosives and entered that way. A U.S. engineer bulldozer captured by the Somalis was destroyed from the air by antitank missiles. Cited in Montgomery P4.

42 In a striking example, the commander of the Italian contingent opened a secret dialogue with Aidid which resulted in a UN request for his relief. As he had acted on the directions of his government, it was denied. "Somalia operations lessons learned," *National Defense University*, October 1, 2002, p. 14.

43 Meade P4, p. 8.

ran counter to the prevailing trends of policy. Montgomery resubmitted a scaled down version, now asking for a reinforced company of Bradley Fighting Vehicles and tanks. This time Hoar agreed to pass the request to the Pentagon.

To their credit, the Joint Chiefs of Staff recommended approval and the chairman forwarded the request to Secretary Aspin. The public record does not show that the military leadership pressed hard, however, and, given the administration's clear intent at the time to downsize the U.S. presence and hand off the mission altogether to the UN—the hunt for Aidid notwithstanding—Aspin's decision to deny Montgomery's request was perhaps predictable.

Even as TF Ranger pursued its search for Aidid, other diplomatic venues were being explored. One involved an attempt to open a channel to Aidid using former president Jimmy Carter, who supposedly enjoyed a previous "relationship" with Aidid and had volunteered to act as an intermediary. Although a legitimate policy initiative, this approach was never communicated to the military leadership in Washington, at CENTCOM headquarters in Tampa, or in Mogadishu. Whether Aidid would have agreed to give up his aspirations to lead Somalia is doubtful; his most likely motives were to buy time, tone down the American pressure and wait for the inevitable U.S. withdrawal. In any case, the Carter initiative died stillborn. Something was about to happen that would change everything.

The Battle of the Black Sea

Mark Bowden's best-selling *Black Hawk Down* (1999), later adapted into an action movie by Ridley Scott, brought the intimate details of October 3 to a national and even global audience. The day began with reports that a number of key Aidid lieutenants planned to meet at the Olympia Hotel, not far from the Bukhara arms market on Hawlwadig Road. Repeating the mission profile that had been used several times previously, TF Ranger launched 160 SOF soldiers (Rangers, Delta operators, SOF aircrew and a small number of SEALs and Air Force para rescue specialists) in sixteen helicopters and twelve vehicles at 3:30 in the afternoon. (Approximately 110 were inserted by helicopter.) Contrary to some reports, only cursory notification—not preliminary coordination—took place between TF Ranger and UNISOM II or the QRF.[44] General Garrison notified General Montgomery of the raid as it was being launched, leaving no opportunity for joint mission rehearsals, exchange of communications plans, or discussion of relief operations or link up procedures under fire.[45]

Confident that the mission would be over in an hour, normal mission essential equipment like night vision goggles, body armor, and even water was in many cases left behind.[46] Although operating on the same tactical battlefield, both the Rangers and Delta Force maintained separate chains of command, with the senior Delta officer aloft in a command and control aircraft and the senior Ranger commander (Lieutenant Colonel Dan McKnight) in charge of the ground vehicle convoy. On the objective, a Ranger captain and Delta captain commanded their respective elements, but neither was designated as the on-scene ground commander. General Garrison exercised overall command from his operations center at the airfield.

Although Somali lookouts reported the launch of the aircraft carrying the raid force, the operation went according to plan until a 160 SOAR Black Hawk, call sign "Super 61," was shot down about fifty minutes into the mission.[47] (The Somalis fired volleys of rocket propelled grenades

44 MG Garrison claimed in a handwritten letter to Congressman Murtha after the battle that he had coordinated for QRF support with MG Montgomery before the raid. Montgomery disputes this in an unclassified after action report on file at the U.S. Army War College, saying he was notified only minutes before the aircraft launched. The confusion and delay evident in organizing the relief force strongly supports Montgomery's version.

45 Information provided to the author by a staff officer personally present in the UNISOM II headquarters at the time.

46 Bowden, p. 39.

47 This aircraft, call sign "Super 61," crashed at 4:40 p.m., 300 meters east of the Olympia Hotel. Both pilots were killed. Three Delta snipers and the two crew chiefs were badly injured. Mark Bowden, *Black Hawk Down* (New York: Penguin Books, 1999), p. 78.

or RPGs at low flying aircraft throughout the battle with great success, especially against the larger and less nimble Black Hawks.)[48] This event disrupted the orderly extraction of the Somali detainees and gave time for Aidid's militia forces and hundreds of angry, armed civilians to flood into the area. Shortly thereafter a second MH60 ("Super 64") was shot down. The lone Combat Search and Rescue (CSAR) helo, "Super 68," was able to insert its medics and Ranger security force at the first crash site but was badly damaged by RPG fire and returned to base. (There was no viable preexisting plan to react to a second downed aircraft.)

The raid now became a full-fledged battle, later dubbed "The Battle of the Black Sea" by the SNA. The ground vehicle convoy carrying the captured SNA leaders, led by the Ranger battalion commander, attempted to respond but came under intense close range fire without reaching the second crash site and was forced to return to the airfield with many dead and wounded. A second, smaller Ranger column then moved out from the airfield in vehicles but was beaten back not far from its start point. At this point, one rifle company from Montgomery's QRF was moved to the American-held airfield and attempted to relieve the embattled SOF troopers but could not advance in the furious city fighting and returned to base.[49] Several hours into the mission, TF Ranger found itself clustered around the two crash sights or pinned down inside several buildings along Marehan Road, unable to disengage from the swarming Somali militia and civilian crowds and unwilling to withdraw without the bodies of their comrades in the downed aircraft.

Unquestionably, the SNA militia and the armed civilian irregulars who participated in the battle were underrated by General Garrison and his special operations staff officers and commanders. Although poorly

48 The previous January, reports reached UNITAF and USLO of large shipments of RPGs entering the country, sent by "Islamic fundamentalists." These accounts were largely discounted as unconfirmed.

49 This unit, C Company, 2nd Battalion, 14th Infantry was ambushed on the Via Lenin not far from the airfield and expended more than 60,000 rounds in half an hour before withdrawing. See Rick Atkinson, "Night of a thousand casualties," *The Washington Post*, January 31, 1994, A1.

equipped and disciplined to American eyes, many were hardened by years of combat. Their ability to mass quickly and fight in large numbers with determination and courage had been amply demonstrated in the days and weeks preceding the October 3 raid. The local SNA commander, Colonel Sharif Hassan Giumale, had trained for three years in Russia and later in Italy, fought in the Ogaden against Ethiopia and commanded a brigade in the Somali National Army before joining Aidid during the civil war.[50] A number of his subordinates were similarly experienced. Well equipped with RPGs and small arms, they had noted the American tactical pattern and its weaknesses. And they were fighting in their own neighborhoods, in front of their families and their clan leaders. Their effectiveness would be grudgingly admitted after the fight, if not before.

At this point, near sundown, the survival of the raid force was very much in question. Dozens had been killed and wounded, at least two separate rescue attempts had failed, more armed Somalis were arriving by the hour, and ammunition was running dangerously low.[51] Of the seven troop-carrying Black Hawks available, five were no longer flyable.[52] Several special operations soldiers died in the field because medical evacuation by air or ground was impossible. Although Aidid's fighters had suffered serious losses, they maintained relentless pressure on the Americans through the night. By most accounts, only the dauntless actions of the AH6 Little Bird pilots, flying all night long, kept the besieged Americans alive through the night.[53]

50 Atkinson, "The raid that went wrong," *The Washington Post*, January 30, 1994, A1.

51 Only heroic low-level resupply runs by a 160 SOAR helicopter, call sign "Super 66," enabled the ground force to keep fighting. Several crew members in this aircraft were badly wounded. In the earlier attack on the Pakistanis on June 5, Aidid's militia had forced them to expend all their ammunition as a tactic, only moving in to slaughter the survivors after they were helpless.

52 Of the eight MH60 Black Hawks used in the raid, two were assigned to insert the Delta snatch team (about thirty personnel), four to insert the Ranger blocking force, one was designated the CSAR (Combat Search and Rescue) aircraft, carrying fifteen personnel plus crew, and one was a specially configured C2 (Command and Control) aircraft. In the course of the battle, five of the seven MH60s capable of carrying troops were shot down or rendered unflyable by enemy fire. Personal memoir by Captain Gerry Izzo, former 160 SOAR MH60 pilot, who participated in the battle.

53 Each AH6 reloaded six times during the night; all eight pilots were awarded the Silver Star. Izzo narrative.

As night fell, General Garrison concluded that the survival of the force was at risk and requested assistance from UNISOM II. Over four hours, U.S. liaison officers worked feverishly to coordinate a rescue force consisting of Malaysian armored personnel carriers, Pakistani tanks, and two companies from the QRF infantry battalion of the 10th Mountain Division. The seventy vehicle rescue force, accompanied by special operations personnel from Garrison's headquarters and TF Ranger support units, moved out at 11:15 p.m. and painfully fought its way to the encircled Rangers and Delta operators, reaching them at 1:55 a.m.[54]

Most of the survivors were wounded at this point. Moving in vehicles and on foot, and carrying their dead and wounded, the dazed Americans retreated to a soccer stadium just outside the combat zone as dawn broke over Mogadishu. Though they had fought hard to recover their dead, the bodies of Shugart and Gordon, as well as the dead aircrew and Delta passengers of Super 61 and Super 64, remained behind.[55] Of the TF Ranger troops who had come to Somalia and entered the fight, seventeen were dead. 106 were wounded.[56] The Rangers were particularly hard hit, with almost every participant killed or wounded. It was, as the British say, a hard day for the Regiment.[57]

Although General Garrison attempted to portray the mission as a success on the grounds that the targeted SNA leaders had been captured, the raid quickly came to be seen as a military and political fiasco.[58] Almost

54 Atkinson, "Night of a thousand casualties," A1.

55 Durant's co-pilot in Super 64, CW3 Ray Frank, survived the crash but was apparently killed later. The Delta operator and two crew chiefs in the rear of the aircraft were either killed on impact or in the fighting which followed. The pilot and co-pilot of Super 61, CW4 Cliff Wolcott and CW3 Donovan Briley, were killed on impact; their bodies were cut out of the wreckage with great difficulty and brought out by the Rangers. The two Delta operators in this aircraft were badly injured and evacuated by a lone MH6 which landed next to the crash site. One of the Delta troopers subsequently died. Bowden, pp. 143, 201.

56 One TF Ranger soldier was killed in a previous raid, and twenty-three were wounded. The statistics normally cited, eighteen dead and eighty-three wounded for October 3–4 include one 10th Mountain soldier but do not include several other 10th Mountain casualties.

57 Somali leaders put their losses at 314 dead and 814 wounded. Atkinson, "Night of a thousand casualties." Atkinson traveled to Mogadishu after the battle and personally interviewed SNA leaders.

58 Garrison's exact words were "the mission was a success. Targeted individuals were captured and extracted from the target." Garrison personal letter to Congressman Murtha, reproduced in Bowden, *Black Hawk Down*, p. 338.

immediately, the Clinton administration came under fierce criticism. Even as a heavy mechanized force was quickly sent in to stabilize the situation, TF Ranger departed, and the hunt for Aidid was quietly dropped. The following spring, U.S. forces pulled out of Mogadishu for good.

Post Mortem

The causes of failure in Mogadishu were not apparent only in hindsight. In many cases, they were fundamental, even blatant; they could, and should, have been identified in advance. Military and civilian leaders in decision making positions bear a heavy share of responsibility for a flawed and ultimately failed policy and for the unnecessary deaths on all sides that resulted. Our tragic experience in Somalia provides critical lessons for military and civilian leaders who bear similar responsibilities for planning and conducting contingency operations now and in the future.

At the political and strategic level, the Clinton administration failed to provide specific, coherent goals and objectives that could be translated into concrete tasks and missions on the ground in Somalia. If the policy objective was "the restoration of an entire country," then the trust and confidence placed in the UN was misplaced, while the resources provided by the U.S. were manifestly inadequate. In particular, the decision to disarm the clans, beginning with Aidid, was pregnant with consequence. It forced the U.S. and UN to abandon the neutrality that had helped make Restore Hope successful, at a time when American military power was growing weaker every day. And it drew the modest U.S. forces in Somalia into high intensity combat operations for which they were not prepared or equipped. The June 5 slaughter of the Pakistanis may or may not have been planned in advance, but the battle lines had been drawn between Aidid and the U.S. well before then. Whatever options applied before that date went up in smoke as soon as the extent of the tragedy became apparent. UNISOM II now faced only two choices: to retaliate by taking down Aidid or to get out of Somalia.

Inside the Beltway, an air of detachment prevailed. No real attempt was made to secure congressional or popular support, an oversight that caused immediate policy failure when casualties mounted. Requests for forces from field commanders were airily dismissed. Long on rhetoric and short on detail, easily distracted by the pressures of domestic politics and other foreign policy challenges and opportunities, the Clinton national security team lost focus on perhaps the most dangerous foreign policy issue then in play.[59] There was a ground truth about conditions in Somalia, waiting to be grasped. The military commanders there saw it clearly. But somewhere between the gutted U.S. compound in Mogadishu and the West Wing, that reality evaporated.

At the operational level, the command relationships established to control forces in Mogadishu proved almost tragicomic. The Commander-in-Chief, U.S. Central Command in Tampa exercised operational control ("OPCON") of two separate combat forces, Garrison's TF Ranger and the 10th Mountain Division's Quick Reaction Force. Those threads came together only in Tampa. No command relationship existed between the two, though they were located five minutes apart.[60] The Commander of U.S. Forces in Somalia exercised no operational authority over any combat forces; at best, he could "borrow" the QRF for short periods, subject to CENTCOM's approval.[61] The failure to designate one officer to command U.S. combat forces in Mogadishu stemmed from the desire of the combatant commander to remain "in charge" and contributed directly to the loss of life in the battle of October 3 and 4.[62] The presence of two major generals, each commanding no more than a few hundred combatants, in

59 Incredibly, Somalia was never the subject of a Principal's Committee until after the fateful events of October 3. Norton, p. 175.

60 Interview with a U.S. staff officer assigned to the COMUSFORSOM command group in Mogadishu at the time.

61 Montgomery was only allowed to exercise "TACON" or tactical control over the QRF, defined as "limited, short duration control for specific local operations." In his official AAR, on file at the Army War College, he was diplomatic but pointed in criticizing this decision.

62 Official joint doctrine at the time was unequivocal: "Command is central to all military action, and unity of command is central to unified action. Unity of command means all forces operate under a single commander with the requisite authority to direct all forces employed in pursuit of a common

the same city during the same ferocious engagement, and linked by little more than their good intentions, predictably caused confusion and delay.[63]

Operational level planning and the resources made available based on it were also badly flawed. As General Mead clearly pointed out, the situation in Mogadishu in September had dramatically changed for the worse. The U.S. forces present in Mogadishu were too small and too lightly armed for the mission. General Montgomery's request for heavy reinforcements lends support to this assessment, as does the urgent decision to send them in force days after the battle.[64] General Garrison's request for return of the AC130 gunships is a similar case in point.[65] U.S. forces manifestly required reinforcement, yet military leaders in the chain of command failed to make a vigorous case—with painful and damaging results.

Tactically, special operations forces in Somalia, lacking context and situational awareness, suffered from over-confidence (Mead's

purpose." In recent years, "Unity of Command" has been replaced in official Joint doctrine with "Unity of Effort, reflecting the inability of the Joint community to agree on unified command. See Joint Publication 0-2, Unified Action Armed Forces.

63 The refusal to place SOF and conventional forces under the same local commander has its roots in the assumption that conventional commanders are prone to misuse highly capable but fragile SOF organizations, or cannot be trusted to maintain operational security. The establishment of the Joint Special Operations Command after the failed Iran rescue mission, the creation of special operations career fields in the services in the 1980s and the advent of U.S. Special Operations Command in 1987 have powerfully reinforced this "rule." For at least a generation, local SOF headquarters have either commanded associated conventional units, as in the Liberia noncombatant evacuation in 1996, or more often remained separate and apart while reporting directly to the regional combatant commander. These firewalls have remained in place despite ample evidence that "stove piping" SOF organizations inhibits battlefield coordination and sharing of information and intelligence.

64 Some critics have asserted that heavy forces could not have arrived in time to intervene in the October 3 fight. See Norton, p. 179. However, the ten days which elapsed between Aspin's "no" decision and the battle were ample to airlift up to two tank companies into Mogadishu. As only four obsolete Pakistani M48 tanks were used in the successful relief operation, this force would have made an enormous difference.

65 After the fact, Garrison argued that AC130 Spectre gunships and armor would have had little effect on the outcome of the operation. (Bowden, p. 340) This is hard to reconcile with the facts. Without the UN armor eventually employed U.S. casualties would have been much higher, and organic US armor would have been much more responsive and effective. Given the exploits of the AH6 gunships, the vastly more capable AC130 gunship, with its much greater loiter time, weapons load, night sensing systems, survivability (due to a much higher operating altitude), massive firepower, and extreme precision, might well have proved decisive in enabling timely extraction of the detainees and friendly forces with minimal loss. This is precisely why Garrison requested them in the first place.

communication shows fairly clearly that the conventional force did not). Virtually all of the advantages possessed by the U.S. military were thrown away: a small force went into a massive urban area, in daylight, without surprise, against greatly superior numbers, without adequate fire support, good intelligence or a strong reserve. Under these conditions, a well-trained, well equipped U.S force with a clear technology overmatch fought at every disadvantage, suffered appalling losses, and came close to annihilation. These risks were run, not because hard intelligence had located Aidid, but to attempt the capture of a few mid-level subordinates. Many tactical errors were fundamental. The failure by TF Ranger to adequately brief and rehearse the 10th Mountain QRF;[66] the decision by small unit leaders to leave behind mission essential equipment;[67] the bifurcated command relationships both inside and outside TF Ranger (which ensured that even individual soldiers fighting in the same room reported to different leaders from different organizations); the repeated use of the same mission template, which allowed the enemy to learn and adapt to American tactics;[68] poor operational security which telegraphed the start of the raid; the use of fragile and thin skinned helicopters at low level over the city in daylight;[69] the failure to plan for the loss of multiple aircraft (not unlikely given the mission profile); the very poor intelligence picture on the capabilities and intentions of the SNA; and the hesitation shown in requesting immediate assistance from the UN all reflect poorly on the commanders involved in planning and executing the raid. The American soldiers who fought the Battle of the Black Sea deserve every accolade bestowed on them. But, they paid dearly for such glory.

66 "COMUSFORSOM was not a factor in planning and evaluating U.S. Ranger operations, even though he was ultimately responsible for reinforcing or extracting the Ranger force . . ." "Report on Operations in Somalia," *U.S. Army Center for Army Lessons Learned*, Fort Leavenworth, Kansas, 2002, p. 3.

67 Bowden, p. 39.

68 Atkinson, "*The Raid That Went Wrong*", A1.

69 On the night of September 25, a U.S. Black Hawk had been shot down over Mogadishu by an RPG with the loss of three aircrew. The nature of the threat to rotary wing aircraft was thus clearly established. Bowden, p. 61.

The lessons of Somalia are hard, but they are clear. Political leaders must be unambiguous about defining the mission and the conditions for success. Congressional and public support are important and deserve effort and attention. The means provided must be sufficient to the task, in size and capability. Multiple, competing chains of command don't work; a single joint commander should be empowered to conduct operations and trusted, not second-guessed. Senior commanders an ocean away cannot control local tactical operations and should not try. Finally, the soldier on the ground in contact with an enemy deserves every advantage America can provide.

The biggest lesson from Somalia is also the simplest. The fight that took place on October 3, 1993, in Mogadishu was a small unit action, a local tactical operation like the several that preceded it. But its effects were devastating, to the administration, to the nation, and to American foreign policy. Whenever American soldiers go in harm's way, they carry America's prestige and credibility with them. If they fail, America's enemies are emboldened and empowered. American power and influence can suffer dramatically for years to come, with impacts that reach far beyond the original mission or policy. America saw that on Marehan Road, now many years ago. We ought not take that road again.

12 Offshore Control

Thinking about Confrontation with China

Avoiding conflict with China is acknowledged as supremely important, yet strategy has been slow to catch up with policy, as seen in the ongoing debate about "anti-access/area denial." Here the authors critique "AirSea Battle," pointing out its inherently escalatory potential, and suggest alternatives that promise better deterrence and better outcomes should deterrence fail.[1]

The rise of China, attended by a more muscular military posture and economic status, has altered the international system in ways that directly challenge America's traditional role in the Asia Pacific region. Prior to 2000, the U.S. enjoyed unrivalled status as the guarantor of stability in the region. Today, the picture is very different. America's persistent budgetary woes, two inconclusive military conflicts, and a bitterly divided U.S. political system give the perception of a declining America. Yet, China may have overplayed its hand and provided the United States with an opportunity to strengthen its position in the Pacific, as China's actions since 2007 have moved Japan, the Philippines, Vietnam, and Korea closer to the United States. China has also stimulated the Association of Southeast Asian Nations (ASEAN) to question China's assertion of a peaceful rise.

1 An abridged version of this paper appeared as "America's ultimate strategy in a clash with China," with T. X. Hammes, *The National Interest*, June 10, 2014.

The Obama administration's principal response, the "Rebalance to Asia,"[2] was intended both to reassure allies and partners in the region and communicate America's enduring interests and role as a Pacific power. For obvious reasons, the administration has been careful not to identify China explicitly as a security threat. Both sides are well aware that, historically, rising and dominant powers often clash. Yet, history also suggests that confrontation with China is far from inevitable. The United States and its allies will not provoke, and would certainly go to extreme lengths to avoid, armed conflict with China.[3] While no set of actions can guarantee continued peace between China and the United States, carefully-considered national and military strategies can reduce the probability of a conflict and, should conflict occur, limit and constrain its fallout and consequences.

Tom Donilon, then the national security advisor, clarified and reinforced the administration's determination to continue its rebalance to Asia in 2012.

> To pursue this vision, the United States is implementing a comprehensive, multidimensional strategy: strengthening alliances; deepening partnerships with emerging powers; building a stable, productive, and constructive relationship with China; empowering regional institutions; and helping to build a regional economic architecture that can sustain shared prosperity.[4]

The United States has clearly articulated the incentives it will use to encourage peaceful growth in the region. While there is an ongoing discussion about how well the Obama administration has executed the diplomatic, economic, and informational aspects of its rebalance, the

2 Remarks by President Obama to the Australian Parliament, November 17, 2011.

3 *National Security Strategy of the United States*, May 2010, The White House, p. 43.

4 Remarks by Tom Donilon, National Security Advisor to the President, "The United States and the Asia-Pacific in 2013," *White House Press Office*, March 2012.

intent is clear. However, to date, the administration has not yet fielded a coherent military strategy that will deter China and reassure U.S. allies and friends in the region.

A major conflict between a rising-China and the United States, Japan, or India is unlikely. Yet, the First World War painfully demonstrates that even countries that are closely integrated economically can clash. It is important that the United States and its friends in Asia work hard to avoid any such confrontation. Further, the U.S.-U.S.S.R. experience indicates that conflict is, in fact, a choice and can be deterred. A key part of avoiding that conflict was the deterrent value of NATO's military forces and a clearly-stated military strategy for resisting Soviet aggression.

AirSea Battle

For this reason, the concept of "AirSea Battle" (ASB) should deeply concern serious observers. Rolled out in 2010 by the Center for Strategic and Budgetary Assessments (CSBA), *AirSea Battle: A Point-of-Departure Operational Concept* postulated that in the "unthinkable" case of a war with China, U.S. forces should attack Chinese surveillance systems and its integrated air defense system, followed by a weighted campaign to bomb Chinese land-based ballistic and anti-ship missile systems to "seize and sustain the initiative in air, sea, space and cyber domains."[5] Though proposed as an operational concept and not a strategy, CSBA provided a blueprint for the development of a new generation of naval and air weapons systems.

AirSea Battle created immediate controversy. Critics fell into two major categories: those who saw it as needlessly provocative,[6] and those who believed it was simply a justification for the Navy and Air Force to

5 Jan van Tol, "Air sea battle: A point-of-departure Operational Concept 2010," *Center for Strategic and Budgetary Assessment*, May 18, 2010.

6 Matt Durnin, "Battle plans tempt chill in U.S.-China relations," *The Wall Street Journal*, November 10, 2011; Daniel Hartnett, "Air-sea battle: Unnecessarily provoking China?" *Center for International Maritime Security*, February 22, 2014; and Amitai Etzioni, "Air-sea battle: A dangerous way to deal with China," *The Diplomat*, September 3, 2013.

gain a greater portion of the defense budget in a time of fierce competition for declining military funding.[7]

For a number of reasons, the CSBA concept is both provocative and would likely prove ineffective. While "blinding" Chinese space-based and ground surveillance systems may make sense in the event the People's Republic of China initiates hostilities, it is dangerous to assume such a campaign will be successful in a time of aerostats, cheap drones, and other emerging technologies such as cube satellites.[8] Further, a weighted air and naval campaign to attack Chinese integrated air defense systems and land-based missile systems is flawed from multiple perspectives. First, it is dangerously provocative. China's Second Rocket Artillery Corps, which controls its conventional land-based missiles, also controls its land-based nuclear arsenal. A direct attack on the organization that controls China's strategic nuclear forces in a scenario where U.S. territory and nuclear forces have not been attacked could escalate the conflict uncontrollably. In this regard, though touted as an "operational concept," AirSea Battle as expressed in the CSBA Concept paper intrudes forcefully and directly into the political domain.

ASB is also questionable because it purports to send a limited number of extremely expensive U.S. assets directly against Chinese strength—its dense and capable air defense network. The concept also counts on the ability of U.S. forces to successfully find and destroy Chinese mobile missile systems, a major failure in Operation Desert Storm in 1991. According to the *Gulf War Air Power Survey,* the coalition air forces saw forty-two launches but could only get into position to drop ordnance eight times.[9] The authors offered in mitigation that commercial vehicles on the highways provided significant background clutter that made the Scuds hard to target. However, the British Special Air Services reported

7 Sydney J. Freedberg, Jr., "Air-sea battle is more about bin Laden than Beijing," *Breaking Defense,* July 16, 2013.

8 James Dorrier, "Tiny cubesat satellites spur revolution in space," *SingularityHUB*, June 23, 2013.

9 William Rosenau, "Special operations forces and elusive enemy ground targets: Lessons from Vietnam and the Persian Gulf War," *The RAND Corporation*, December 31, 2001, p. 34.

that actual launches could be seen by ground observers from thirty miles away.[10] In addition, Allied forces had absolute air supremacy, as well as hundreds of aircraft that could range freely over the entire country. Despite all these advantages, the *Survey* concluded that "There is no indisputable proof that Scud mobile launchers—as opposed to high-fidelity decoys, trucks, or other objects with Scud-like signature—were destroyed by fixed-wing aircraft."[11] Despite a massive effort involving thousands of air sorties, ground and national intelligence assets, the allies failed to get a single confirmed kill, even though it took the Iraqis at least thirty minutes to erect, fuel, and launch a liquid fueled Scud in relatively open desert terrain.[12] It is possible, but not likely, that airpower will fare better against the solid fueled, much more numerous Chinese systems in the complex, heavily-defended environment of coastal China. These mobile systems can be hidden in garages, buildings under construction, caves, and tunnels. There are tens of thousands of places to hide. The launch vehicles can also be camouflaged as commercial vehicles for the periods when they move between hiding places. Finally, solid fueled systems can launch in much less time than liquid fueled ones and can be hidden in garages, Thus, to succeed, the United States would have to maintain enough aircraft in the contested airspace of China to detect the missile and place a weapon on it in a matter of minutes.

In short, the CSBA AirSea Battle concept has very little chance of success in any of the three areas it considers vital—blinding, eliminating command and control nodes, or suppressing launch systems. Yet, it accepts the high risk of attacking mainland China and the probability that such an attack will make conflict resolution even more difficult.

These deficiencies clearly call into question ASB's deterrence value against China. Because ASB apparently depends upon space and cyber

10 Rosenau, p. 38.

11 Thomas A. Keaney and Eliot A. Cohen, "Gulf War air power survey summary report," *The RAND Corporation*, 1993, pp. 89–90.

12 Rosenau, p. 32.

systems, China may well feel it can degrade those systems enough to defeat the operational approach. Further, China may believe the United States cannot afford ASB or, at very least, will not field the capabilities for a decade or more. A military concept that is vulnerable to a relatively inexpensive defeat mechanism or has a window of vulnerability has little deterrent value.

Perhaps the most strategically significant weakness of AirSea Battle is that it may frighten our allies as much or more than our potential adversary. Since much of the technology in DOD's actual ASB program is top secret, U.S. officials are unable to discuss it with our allies. As a result, many allies assume it will follow the pattern described in CSBA's paper and initiate immediate, extensive attacks on Chinese territory. They are obviously concerned that China will see such attacks as emanating from allied territory and respond in kind. U.S. allies are, in effect, being asked to provide bases without any knowledge of what actions the U.S. intends to take from those bases.

Since the publication of CSBA's paper, the Pentagon has repeatedly stated that it does not represent U.S. policy. Both the Chief of Naval Operations and the Chief of Staff of the Air Force have stated ASB is not a strategy and is not directed at China.[13] However, the Joint Staff has also published the Joint Operational Access Concept (JOAC) that provides the doctrine for gaining access in an anti-access/area denial (A2/AD) environment—and it uses many of the ideas from the ASB paper. At the same time, the Pentagon has created an AirSea Battle Office, not as an operational concept but as "a help desk for the A2/AD fight."[14] One can forgive our allies if they are somewhat unclear on exactly what ASB is.

The absence of any other publicly discussed military strategy for a potential conflict with China means many people, including senior foreign officials, believe AirSea Battle, as expressed in the CSBA paper,

13 Jonathan Greenert and Mark Welsh, "Breaking the kill chain," *Foreign Policy*, May 16, 2013.

14 Sam Lagrone and Dave Majumdar, "The future of air sea battle," *US Naval Institute News*, October 31, 2013.

remains the U.S. strategy. In the absence of any stated U.S. military strategy, CSBA's concept will continue to fill the vacuum.

A Proposed Military Strategy

The U.S. commitment to peace and avoiding conflict with China is genuine and unambiguous. Keeping that peace depends, in part, on an affordable military strategy that goes well beyond "operational concepts" to address China's growing military and economic strength and its clear intent to extend its power and presence in Asia. Any U.S. military strategy for Asia must achieve six objectives:

- To deter China from military action to resolve disputes while encouraging its continued economic growth
- To ensure access for U.S. forces and allied commercial interests to the global commons
- To assure Asian nations that the United States is both willing to and capable of remaining engaged in Asia
- To discourage friends and allies from taking aggressive steps that further destabilize the region
- In the event that deterrence fails, to achieve U.S. objectives with minimal risk of nuclear escalation
- To be visible and credible today, not years in the future.

In the absence of any published military strategy, we propose "Offshore Control: Defense of the First Island Chain" as an effective and affordable approach for a conventional conflict with China. Offshore Control establishes concentric rings that *deny* China the use of the sea inside the first island chain, *defend* the sea and air space of the first island chain nations, and *dominate* the air and maritime space outside the island chain. Offshore Control does not strike into China but takes advantage of geography to block China's key imports and exports and, thus, severely weaken its economy. No kinetic operations will penetrate Chinese airspace. Prohibiting penetration is intended to reduce the possibility of nuclear escalation and make conflict termination easier.

This approach would exploit China's military weaknesses, which increase exponentially beyond the "first island chain" running through the Japanese archipelago, the Ryukyu Islands, Taiwan, the northern Philippines, and Borneo to the Malay Peninsula. Allied naval and air forces attempting to operate near or on Chinese territory face daunting odds. In contrast, allied forces fighting as part of an integrated air-sea-land defense of the first island chain gain major tactical advantages over Chinese forces. Outside that arc, Chinese capabilities dwindle markedly.

Denial as an element of the campaign plays to U.S. strengths by employing primarily attack submarines, mines, and a limited number of air assets inside the first island chain. This area will be declared a maritime exclusion zone with the warning that ships in the zone will be attacked. While the United States cannot stop all sea traffic in this zone, it can prevent the passage of large cargo ships and large tankers, severely disrupting China's economy relatively quickly. As an integral part of denial, any Chinese military assets outside the Chinese twelve-mile limit will be subject to attack.

In most foreseeable scenarios, China would act alone. If the United States manages its alliances well, it would not. The *defensive* component of Offshore Control will exploit this advantage to bring the full range of U.S. and allied assets to defend allied territory and encourage allies to contribute to that defense. It exploits geography to force China to fight at long range while allowing U.S. and allied forces to fight as part of an integrated air-sea-land defense over their own territories. In short, it will flip the advantages of anti-access/area denial (A2/AD) from China to the allies. Numerous small islands from Japan to Taiwan and on to

Luzon provide dispersed land basing options for air and sea defense of the apparent gaps in the first island chain. Since Offshore Control will rely heavily on land-based air and sea defenses, to include mine and countermine capability, we can encourage potential partners to invest in these capabilities and exercise together regularly in peacetime. The United States will not request any nations to allow the use of their bases to attack China. The strategy will only ask nations to allow the presence of U.S. defensive systems to defend that nation's air, sea, and land space. The U.S. commitment will include assisting with convoy operations to maintain the flow of essential imports and exports in the face of Chinese interdiction attempts. In exercises, the United States could demonstrate all the necessary capabilities to defend allies—and do so in conjunction with the host nation forces.

The *dominate* phase of the campaign will be fought outside the range of most Chinese assets and will use a combination of air, naval, ground, and rented commercial platforms to intercept and divert the super tankers and very large container ships essential to China's economy. Interdicting Chinese energy imports will weaken but not destroy China's economy. China can and has taken steps to reduce the impact of an energy blockade. Fortunately, exports are of even greater importance to the Chinese economy. Those exports rely on large container ships for competitive cost advantage. The roughly 1,000 ships of this size are the easiest to track and divert. Naturally, China will respond by rerouting, but all shipping to China must pass through the first island chain. Even if China seizes a portion of the chain, the United States and its allies can use the more distant choke points of Malacca, Lombok, Sunda Straits, and the routes south of Australia to cut trade from the west. To cut trade from the east, the United States needs only to control the Panama Canal and the Straits of Magellan—or, if polar ice melt continues, the northern route.

While such a concentric blockade campaign will require a layered effort from the straits to China's coast, it will mostly be fought at a great distance from China—effectively out of range of most of China's military power. The only ways for China to break the blockade is to build a global sea control navy or develop alternative land routes. A sea control navy will require investing hundreds of billions of dollars over decades. Alternate overland routes simply cannot move the 9.74 billion tons of goods China exported by sea in 2012.[15] This is the equivalent of roughly 1,000 trains per day each way over the two rail lines that link China to Europe.

Further contributing to Offshore Control's credibility is the fact the United States can execute the campaign with the military forces and equipment it has today. Unlike CSBA's *AirSea Battle Concept*, it does not rely on expensive, highly classified, developmental defense programs for success. Rather, the United States can exercise the necessary capabilities with its allies now, demonstrating that today's capabilities are sufficient to execute the strategy.

This brings us to the ends the strategy seeks. Offshore Control is predicated on the idea that the presence of nuclear weapons makes decisive victory over China too dangerous to contemplate. The United States does not understand the Communist Party's decision process for the employment of nuclear weapons, but it does know the Party will use all necessary means to remain in power. Instead of attacking China where it is strongest, Offshore Control seeks to use a war of economic pressure to bring about a stalemate and cessation of conflict with a return to a modified version of the status quo. Faced with the threat of economic collapse, China's leaders are far more likely to bargain. Theoretical strategists may question the lack of a path to decisive victory, but decisive victory falls outside the logic of conflict with a great nuclear power. There, one seeks to avoid the clash and, failing that, to achieve acceptable outcomes that enable all sides to back away. In this sense, Offshore Control offers a more realistic and pragmatic roadmap to resolution and peace.

15 "Chinese ports: Throughput up 6.8%," *Shipping herald: The maritime portal*, February 13, 2013.

Conclusion

Perhaps the most important security issue for the United States and its friends in the Pacific is how to encourage China's growth and continuing integration into the world economy while still deterring it from using force to achieve its goals. In short, rather than great power rivalry, we seek to achieve great power coexistence.

President Obama has presented a national strategy that sets those goals and lays out the diplomatic, economic, and political paths necessary to achieve them. However, the United States has yet to articulate a complementary military strategy. Offshore Control is a starting point for a discussion with our allies and friends in the region. It seeks to provide the military component of the U.S. national strategy in Asia. The strategy looks to two major goals in peacetime. The first is to encourage China's economic growth via further integration into the global economy. Obviously, China's continued growth is essential for international economic prosperity. China's continued integration with the global economy also makes Offshore Control potentially more effective, as the more reliant China is on exports, the more vulnerable it is to blockade. Further, by demonstrating it is operationally a defensive approach, we can show the Chinese we do not have aggressive designs on their homeland.

The second major goal of Offshore Control is to deter China by presenting it with a strategy that Chinese strategists know cannot be defeated easily or quickly. This directly addresses one of the most worrying aspects of the current situation in Asia. Like the Germans before WWI, the Chinese may believe they can win a short war. In particular, they may believe their growing capabilities in space and cyber might neutralize U.S. power in the region. By showing that Offshore Control can be executed with today's force, even with dramatically reduced access to space and cyber, the United States and its allies can dispel the notion of a short war. Strengthening this approach is the fact that the historical record of the last two centuries shows wars between major powers were long—generally measured in years, not weeks or months. A long war means China will have to face the inevitable debilitation of a blockade. The only way China can defeat such a strategy is to create a global sea control navy or develop land routes that are economically competitive with sea routes. Neither will be a guarantee of success. Much of Offshore Control's deterrence comes from the fact it directly addresses two of China's enduring strategic fears—a prolonged conflict and its "Malacca dilemma."

Adding Offshore Control as the military element of the rebalance to Asia provides a military strategy that supports the policy stated by President Obama. It can assure our allies that America has the will and the capability to prevail in a military confrontation. It can deter China by making it clear there will be no easy win in such a confrontation. The primary goal of the U.S. national strategy is to convince China that Great Power Rivalry is a poor choice because the cost of rivalry is simply too high. Far more than AirSea Battle, Offshore Control can serve these ends.

13 Iraq and Afghanistan

Reflections on Lessons Encountered

with Joseph J. Collins

"The Long War"—America's engagement in Iraq and Afghanistan following the 9/11 attacks—evolved into the longest conflict in American history. The end is not yet in sight. In both campaigns, initial successes gave way to protracted, painful, and inconclusive operations that exposed shortcomings in policy, strategy and the conduct of military operations. The costs in blood and treasure have been inordinate, but the gains, while real, have not been proportionate. What have we learned? [1]

This essay attempts to capture, at the strategic level, useful lessons from America's long and painful experience in Iraq and Afghanistan. The task has been daunting, not least because we find ourselves far enough removed from events to lend a measure of clarity, but not so far as to permit true objectivity. This is not history, at least not yet, nor is it revealed truth. But, we know enough now to render an early accounting and to offer for consideration findings and recommendations that may support and inform the soldier/statesman and the strategist when facing similar complex challenges.

Iraq and Afghanistan loom large in the popular consciousness as the long, extended, grinding conflicts that dominated American political life, along with the economic collapse of 2008, in the years following 9/11. Both

1 Excerpted from *Lessons encountered: Learning from the Long War*, National Defense University Press, 2016, (with Joseph J. Collins).

were separate and distinct cases, yet each was inextricably involved with the other, usually as competitors for resources. Both began as more or less conventional state-on-state military interventions but evolved quickly into full-blown counter-insurgencies. Both involved large coalitions, massive security assistance programs, and bitterly divided ethno-sectarian groups, challenging attempts to employ a "comprehensive approach" that could unite civil and military action across the effort. Both featured weak, corrupt host nation governments. Yet there were important differences. Iraq featured greater wealth, a more advanced infrastructure, easier logistical challenges, different tribal and ethno-sectarian dynamics, and more human capital. Afghanistan, lacking oil and other natural resources, was desperately poor and vulnerable to outside intervention, while its harsh climate and topography made military operations difficult. As in Vietnam, the U.S. military was forced to adapt its doctrine, training, and equipment in nonstandard ways, while the civilian component strained to build host-nation capacity.

With this as context, we can say unequivocally that the wars in Afghanistan and Iraq carried high costs in blood and treasure. More than 10,000 American service members, government civilians, or contractor personnel have been killed, and well over 80,000 have been wounded or injured, many seriously. Veterans and service members suffering post-traumatic stress or traumatic brain injury add hundreds of thousands more. Our allies and partners, not including host nations, count over 1,400 dead. In Iraq alone, at least 135,000 civilians were killed, mostly by terrorists and insurgents.[2] In Afghanistan, from 2009 to 2014, nearly 18,000 civilians were killed, over 70% at the hands of the enemy.[3] The effects of these wars, at home and abroad, will be felt for many years to come.

2 Estimates of civilian dead in Iraq vary widely, with the low end at approximately 133,000. See "Costs of war," *The Watson Institute for International Studies*, Brown University, May 2014.

3 UNAMA, *Annual Report 2014: Protection of Civilians in Armed Conflict*, February 2015, pp. 1–2. Accessed from http://unama.unmissions.org/Portals/UNAMA/human%20rights/2015/2014-Annual-Report-on-Protection-of-Civilians-Final.pdf

The direct and related financial costs of these campaigns approach three trillion dollars, which in the main were not covered by revenues but were put on the nation's "credit card," adding to the national debt.[4] The U.S. Armed Forces—especially its ground forces—experienced extraordinary stress and have yet to fully recover. That recovery has suffered from the twin challenges of sequestration and the requirements of new and pressing conflicts.

Fourteen years after 9/11, any attempt to accurately gauge political losses and gains from the wars in Iraq and Afghanistan is problematic. The costs appear high and the benefits slight, though long term outcomes remain uncertain. Iraq, thought to have been stabilized in 2011 when U.S. and coalition troops withdrew, now faces partition and a strong pull into an Iranian orbit. Far from destroying al-Qaeda in Iraq (AQI), the Coalition campaign in Iraq has seen its successor, the Islamic State in Iraq and the Levant (ISIL), further destabilizing Iraq, Syria, and the region as a whole. Afghanistan under the new Ghani administration remains a work in progress, its future after the withdrawal of ISAF in question.

Looking back at this remove, the costs seem clear, painful and excessive. The benefits are unclear or beyond the horizon. Throughout, the U.S. Armed Forces performed with courage and competence, retaining the trust and confidence of the American people. Yet, success in both campaigns is elusive. Progress in Afghanistan and Iraq, in General David Petraeus's words, still appears fragile and reversible.

There have been solid gains. Saddam Hussein's tyranny, aggression, and lust for weapons of mass destruction are history. Al-Qaeda Central in Afghanistan and Pakistan has been all but destroyed. The Taliban

4 Costing Iraq and Afghanistan is imprecise because methodology can vary widely. For example, future interest payments and veteran's care in the outyears is counted in some estimates and not in others. Federal obligations for the Departments of Defense, State, and Veteran's Affairs related to the two campaigns from 2001–2014, including war-related increased to DOD's base budget and war-related aid and assistance to Pakistan but excluding Homeland Security, war-related interest payments and estimated costs for veteran's care through 2054, are $2.6 trillion. If future costs for veteran's care are included, the figure rises well above $3 trillion. If future interest on war-related debt is included, the total exceeds $4 trillion. See Neta C. Crawford, "Summary of costs for the U.S. wars in Iraq, Afghanistan, and Pakistan FY2001–2014," Boston University, June 25, 2014.

have been checked, although their various branches remain a potent force in both Afghanistan and Pakistan. Because of the dedicated work of our intelligence community, the armed forces, the Department of Homeland Security, and the national law enforcement establishment, al-Qaeda has been unable to repeat the catastrophic attacks of September 2001. This is a crowning achievement of the Long War and one that should not be discounted.

Both Afghanistan and Iraq have been liberated from highly oppressive regimes. They have also been introduced to democracy. More immediately, both nations have received generous help in reconstruction. Afghanistan, for example, had been at war for nearly twenty-four years before the United States and its partners helped to oust the backward and highly authoritarian Taliban regime. The devastation of the country in 2002 stands in great contrast to the effects of U.S. and allied reconstruction efforts, which have greatly improved the quality of life for Afghan citizens.[5]

Al-Qaeda terrorism, however, has morphed from a single hierarchical organization to a set of interlocking networks. There are now al-Qaeda rivals, like ISIS, that have significant capabilities, and other violent extremist organizations, especially in North Africa and the Horn of Africa, that have declared themselves to be members or affiliates of al-Qaeda. U.S. Lieutenant General Michael Flynn, the outgoing head of the Defense Intelligence Agency, noted that "In 2004, there were twenty-one total Islamic terrorist groups spread out in eighteen countries. Today, there are forty-one Islamic terrorist groups spread out in twenty-four countries."[6] While we may have prevented major terrorist attacks against the homeland since 9/11, we have no reason to be complacent.

5 Highlights include an increase in adult life expectancy from 42 to 64 years; a decrease in maternal mortality from 1,600% to 327 per 100,000; a more than ten-fold increase in school attendance (including a 36% increase for girls); an increase in one-hour access to health care from 9% to 60%; an increase in reliable access to electricity of 12%; and a growth in mobile phone subscribers from zero to 19 million.

6 James Kitfield, "Flynn's last interview: Iconoclast departs DIA with a warning," *Breaking Defense*, August 7, 2014. Accessed from http://breakingdefense.com/2014/08/flynns-last-interview-intel-iconoclast-departs-dia-with-a-warning/

In geostrategic terms, our intervention in Iraq has accelerated the Sunni-Shia conflict that now rends the Middle East. Saddam Hussein was an odious tyrant, but his Iraq represented a powerful counterweight to Iranian hegemonic aspirations. Iran has been the great gainer—as a weakened and fractured Iraq, dominated by Shia political forces—and is now heavily influenced by the mullahs in Tehran. The intense sectarianism that followed the U.S. departure from Iraq enabled the rise of ISIL in the years that followed. This consequence was unforeseen and unintended.

It is important to note that neither Iraq nor Afghanistan were originally counterinsurgency campaigns. Both interventions unseated existing governments, to be replaced by new leaders, nascent governance structures, and a bewildering array of coalition and international development and aid organizations that flooded these countries with money and advisers. Practical, working democracy, competent ministries, and the rule of law did not materialize quickly, frustrating the desire to hand over governance and security responsibilities and withdraw. In both cases, the opposition was defeated but not destroyed. Over time, strong insurgent forces were reconstituted to contest host-nation governance and coalition security forces. For various reasons, U.S. leaders were slow to acknowledge the nature and character of these conflicts, though our subsequent adaptation was both rapid and effective.

In Afghanistan, a sober assessment shows that, while the Afghan people are clearly better off than they were under the Taliban, and while Afghanistan is no longer a safe haven for al-Qaeda, the Taliban were not eliminated. The future stability and prosperity of Afghanistan remains in some doubt. Pakistan's interest in a weak and destabilized Afghan state has not diminished after many years of partnership and financial support from the United States.

In both campaigns, the U.S. and its coalition partners were ultimately successful in establishing a level of security needed to enable a political settlement. Our military efforts were able to set conditions and create space for a resolution of the political issues which had impelled both insurgencies in the first place. This must be seen as a major accomplishment. Unfortunately, both the Iraqi and Afghan political establishments lacked the will and capacity to exploit these gains. Internal corruption and inadequate democratic structures, grafted onto traditionally authoritarian and tribal cultures, prevented stable, power-sharing political solutions. Thus, the military gains achieved were not alone sufficient to enable political solutions, despite committing huge sums and the sustained efforts over many years of coalition diplomats and development experts. Herein lies a powerful lesson: by itself, the military instrument cannot solve inherently political questions, absent the total defeat of an adversary and its reshaping from the ground up. This is unlikely in all but the most extreme cases.

There is, however, a larger context. The ideological and sociological seeds of Islamist terrorism and insurgency have roots, and these do not lie directly with the West. They are, rather, side effects of a larger war within Islam between fundamentalist and more moderate camps and a struggle for political modernization in a greater Middle East much in need of reform.[7] So long as economic opportunity and dignity are denied to most, violent extremist organizations cloaked in radical interpretations of Islam will likely persist. This suggests that the conflict we have been engaged in for the past fourteen years will continue, albeit in new forms.

The long war has become a longer war; as Clausewitz noted, the results of war are never final.[8] Those who crave a final accounting of the wars in Iraq and Afghanistan will have to wait decades to get it. It is, however, possible to offer judgments and observations that may be helpful to the rising generation of senior military leadership. Both

7 "In the next century few things will matter more than the battle for the soul of Islam; should fundamentalist brands triumph and become mainstreamed, the destabilizing effects throughout the Islamic world and the community of nations itself will be almost incalculable." R. D. Hooker, Jr., "Beyond Vom Kriege: The character and conduct of modern war," *Parameters*, Summer 2005, p. 11.

8 Carl von Clausewitz, (1976). *On War*, pp. 80. Michael Howard and Peter Paret (Eds.). Princeton, NJ: Princeton University Press.

civilian and military leaders are required to cooperate to make effective strategy, yet their cultures vary widely. As noted elsewhere, the dialogue is an unequal one, with the power of decision residing exclusively with the President and the civilian leadership. Nevertheless, the role of senior military leaders is critical. If military professionalism means anything at all, they possess expert knowledge not available anywhere else. By law and precedent, they have a right to be heard. Navigating this terrain represents the art of generalship at its most challenging. Success derives from intellectual preparation, decades of experiential learning and high success in leading complex military organizations, a decided character that is sturdy and self-confident while also open to new ideas, and a strong moral-ethical compass. Not all rise to the top of the military hierarchy so equipped.

While a comprehensive discussion of findings and observations is found in earlier chapters, Operations Iraqi Freedom and Enduring Freedom represent distinct case studies in how policy and strategy are made, each a rich vein to be mined. Immediately following the 9/11 attacks, an urgent consensus formed demanding a military response. In the case of Afghanistan, time was short, and only very limited interagency discussion took place before military forces were in motion. In a sense, our approach to the campaign was always, in Moltke's felicitous phrase, a "system of expedients" as the interagency adapted and evolved to changing conditions and to the reality that, for many years, Afghanistan was a secondary priority to Iraq. Only in 2010 did Afghanistan become the primary theater of war.

The opportunity for planning and preparation was far greater in the case of Iraq. Here, the case for war was less clear, the higher prioritization less convincing, the military less enthusiastic. Perhaps the most basic of strategic questions—what is the problem to be solved?—became a football to be kicked around for the next several years, from destruction of WMD to preventing a nexus of terrorism to establishing democracy in the heart of the Arab world. Many key assumptions—that Saddam's WMD program presented a clear and present danger, that the war would pay for itself, that the majority Shia population would welcome coalition forces, that working through Iraqi tribal structures could be safely ignored, that a small footprint could be successful, that large scale de-Baathification was needful and practical, that a rapid transfer to Iraqi control was possible—proved unfounded, dislocating our strategy and the campaign. The failure to plan adequately and comprehensively for the post-conflict period ushered in a new, dangerous and intractable phase that saw a rapid descent into, first, insurgency and, then, civil war.[9] National decisions to tie strategic success to corrupt and incapable host nation governments—the primary drivers of the insurgencies in the first place—constituted perhaps the most fateful judgments of all. If senior military figures had reservations on these key points, they were not vigorously pressed, a point that will exercise historians for decades.

What was the appropriate role for senior military figures in this regard? The answer lies partly in the degree to which military leaders at the politico-military interface are expected to limit their advice to purely military matters—to delivering "best military advice" only, leaving aside political, economic, legal, and other dimensions for others to weigh. This is a recurring theme in civil-military relations, dating to the 1950s, if not earlier, which has not yet been fully resolved. Political leaders may feel, and some clearly do, that military officers are ill-equipped to operate in this environment.[10]

> [M]ilitary officers are ill-prepared to contribute to high policy. Normal career patterns do not look towards such a role . . . half-hearted attempts at irregular intervals in an officer's career to introduce him

9 "I had trouble that fall getting [General] Franks to focus on Phase IV [post-conflict operations] . . . CENTCOM's planning for Phase IV never improved." General (Ret) Richard B. Myers, *Eyes on the horizon* (New York: Pocket Books, 2014), p. 225.

10 ". . . on politico-military issues . . . military officers may have little background and no strong views of their own." Doug Feith, *War and decision,* (New York: Harper, 2008), p. 371.

> to questions of international politics produce only superficiality and presumption and an altogether deficient sense of the real complexity of the problems facing the nation.[11]

An alternate perspective, voiced by President Kennedy but with roots in Clausewitz, holds that military officers engaged at the highest levels have not only a right but also a duty to take into consideration the context of critical national security issues, including their political, diplomatic, and economic dimensions, lest their military advice be rendered useless or impractical.

President Kennedy specifically urged—even ordered—the military, from the Joint Chiefs right down to academy cadets, to eschew "narrow" definitions of military competence and responsibilities, take into account political considerations in their military recommendations, and prepare themselves to take active roles in the policy-making process.[12]

We take the latter view. For the Chairman, the Joint Chiefs and the Combatant Commanders, there is no "purely military" question, no neat distinction between different dimensions of strategy and policy. They are conjoined. In this regard, we do not find in the record convincing evidence of vigorous debate or respectful dissent from senior military leaders on the key questions raised above, though admittedly all rise above the purely military. Nor do we see them as apparent only in hindsight. The military operations leading to the overthrow of Saddam were outstandingly successful, a tribute to superb military leadership and to the Armed Forces as a whole. Nevertheless, the basic assumptions upon which our national and campaign strategies for Iraq were based were flawed, with doleful consequences. The primary responsibility must lie with the political leaders who made them. But senior military leaders

11 John F. Reichart and Steven R. Sturm, (Eds.), *American defense policy* (Baltimore, MD: The Johns Hopkins University Press, 1987), p. 724.

12 Jerome Slater, "Military officers and politics I," in Reichart and Sturm, *American defense policy*, p. 750.

also have a voice and real influence as expert practitioners in their fields. In the case of the decision to invade Iraq, this influence was not used in full.[13]

This dynamic speaks fundamentally to how we make strategy in America and how our civil-military relations are ordered. Despite criticism of the military as "praetorian" or "out of control,"[14] deference to civilian control is real, especially when dealing with very strong civilian personalities. The example of early success in Afghanistan with limited forces empowered proponents of a similar approach for Iraq, as did the very heavy support by the administration for "transformational" thinking about armed conflict. In terms of organizational culture, and our experience in Vietnam notwithstanding, the Armed Forces were more predisposed to sharp, decisive, conventional operations than protracted irregular ones. These factors help explain, in part, the role played by senior military leaders in the run-up to Iraq.

A separate but related case is President George W. Bush's decision to surge in Iraq in 2006, made against the recommendations of the military chain of command. The ultimate success of the surge remains open to debate. Some argue that the surge precipitated a major reduction in violence, creating conditions for a political settlement that ultimately failed when U.S. forces withdrew in 2011. Others see the crisis in Iraq today as evidence that the surge, along with the awakening, was only a tactical success with temporarily positive effects that were undone later by the political failures of the Maliki administration. While these differing

13 To most observers, Secretary of Defense Rumsfeld was an imposing and forceful personality who did not encourage dissenting views. Few previous secretaries, and perhaps none, constructed decision settings more difficult to work in than his, a perspective widely shared by those outside his immediate circle. However, Rumsfeld himself insists that he encouraged debate and in fact demanded it. Donald Rumsfeld, *Known and unknown* (New York: Sentinel, 2011), p. 456.

14 Historian Richard Kohn is a primary exponent of this view. See Richard H. Kohn, "Out of control: The crisis in civil military relations," *The National Interest, 35*, (Spring 1994); "The Forgotten fundamentals of civilian control of the military in democratic government," John M. Olin Institute for Strategic Studies, *Project on US Post Cold-War Civil-Military Relations*, Working Paper No. 13, Harvard University, June 1997; and "The erosion of civilian control of the military in the United States today," *Naval War College Review, 55* (Summer 2002).

perspectives will not be resolved, the role played by senior military leaders at this time illuminates both the strengths and weaknesses of America's unique approach to making strategy.

2006 was a difficult year for the U.S. and the coalition in Iraq. The February 22 bombing of the al-Askari Mosque (the "Golden Dome" in Samarra, sacred to Shia Islam) led to an extraordinary spike in violence. According to the Sunni Association of Muslim Scholars, 168 Sunni mosques were attacked within two days of the bombing. By most accounts, Iraq began to degenerate into open civil war, a conflict which the new government of Nouri al-Maliki was unable to control. Several attempts to stabilize Baghdad failed. That summer, officials with the National Security Council staff began to push for a "policy review." In November, the administration was dealt a strong rebuff in the mid-term elections, leading to the dismissal of Secretary Rumsfeld. In early December, the bipartisan Iraq Study Group released its report, saying "the situation in Iraq is grave and deteriorating."[15]

Aware that the success of the campaign was in doubt, President Bush reached out to a number of advisers, both in and outside the formal military and political chains. He was provided essentially with three options: to accelerate the withdrawal of American troops and the handover to Iraqi security forces; to pull back from the capital and allow the factions to fight it out; or to surge forces dramatically to regain the initiative and reestablish security.[16] With some variations, most senior military officials favored the first option. In the case of the Joint Chiefs, their views were undoubtedly colored by their Title 10 responsibilities to preserve a force weakened by years at war, as well as concerns about readiness to meet other contingencies should they erupt. Other senior commanders genuinely felt that more U.S. troops would only inflame local opposition from both sides. In the end, the President elected to

15 James A. Baker III and Lee H. Hamilton, co-chairs, *The Iraq study group report* (New York: Random House, 2006), p. xiii.

16 George W. Bush, *Decision points* (New York: Crown, 2010), p. 372.

surge five Army brigades to the capital and 4,000 Marines to Anbar Province in western Iraq.

In so doing, President Bush chose not to adopt the military advice provided by the formal chain of command, opting instead for the surge option recommended by outside advisers. Moving swiftly, he replaced Rumsfeld with Robert Gates, installed Petraeus as his new field commander, announced an increase in the size of the Army and Marine Corps, directed an associated "civilian surge," and expedited the deployment of the fresh troops. To their credit, senior military leaders supported the President's decision and its implementation, helping to enable a 95% reduction in violence and setting conditions for an eventual transition to Iraqi control. This achievement staved off defeat and a precipitous withdrawal, perhaps the best outcome available under the prevailing circumstances.

Any scholar assessing this period must confront the fact that, in this case, the President as Commander-in-Chief disregarded the best military advice proffered by the JCS, the combatant commander and the theater commander. (To be fair, President Bush encountered opposition from the state department, Congress, and his own party as well.) Plumbing the depths of this paradox requires more space than we have here, but a true understanding has many dimensions. Many of the three- and four-stars engaged in Iraq in 2006 spent most of their careers focused on conventional warfighting and not on counterinsurgency; indeed, the debate on the efficacy and applicability of COIN doctrine continues to this day. Most of them had specific responsibilities and frames of reference that did not encompass the President's very wide field of view. It is also worth noting that, by late 2006, the President had been engaged and focused on Iraq for at least four years and was by then experienced and highly knowledgeable.[17] The recommendations of senior military leaders can be seen as grounded in their particular backgrounds, sets of experiences, and personal perspectives, none of which mirrored the President's.

17 In addition to weekly meetings of the National Security Council on Iraq, President Bush also received and read daily updates produced by the NSC Office for Iraq and Afghanistan.

A fair rendering of this episode might conclude that, at bottom, the system worked as it should. For his part, President Bush was careful to solicit the views and inputs of his most senior military and civilian advisers and weighed them carefully. This give and take was clearly helpful to all concerned. Yet, he also went outside the circle of formal advisers to ensure that all points of view were brought forward. His ultimate decision was clear and unambiguous, and he generously supported the requests of his military commanders. Against strong opposition in Congress and much criticism in the media, he displayed a persistence and determination that proved most helpful to the theater commander and chief of mission charged with implementing his strategy. In his time in office, much went wrong in Iraq, and observers have found much to criticize. By any standard—and the ultimate outcome in Iraq notwithstanding—this decision and its implementation must stand as a high point in President Bush's administration and a successful example of civil-military interaction.

Three years later, President Obama found himself in a similar quandary. For several years, a resurgent Taliban had pressed U.S. and NATO forces. This prompted an increase in troop strength in 2008, bringing the full continent of coalition forces to 68,000. As U.S. troop numbers in Iraq came down, and as the security situation in Afghanistan worsened, the new administration authorized another 17,000 U.S. troops in February 2009 and replaced General David McKiernan in June with General Stanley McChrystal, thought to be a commander with greater skills in counter-terrorism and counterinsurgency.[18] After conducting his own strategic review, McChrystal requested a further 40,000 troops, warning that "failure to provide adequate resources risks . . . mission failure."[19]

This episode provoked serious debate and discussion in the inter-agency and has been widely covered in the memoirs of senior officials.

18 Robert Gates, *Duty: Memoirs of a secretary of war* (New York: Alfred A. Knopf, 2014), p. 346.

19 The president eventually ordered a troop surge of 30,000 as recommended by Secretary Gates, announcing his decision at West Point on December 1, 2009. See Bob Woodward, "McChrystal: 'More forces or mission failure,'" *The Washington Post*, September 21, 2009.

At issue was the split between White House officials who opposed a large increase and military officials who supported it. (Secretary Gates found himself somewhat in the middle, straddling the divide and attempting to manage an increasingly fractious process.)

A deeper question was the approach adopted by senior military officials during policy deliberations. At the time and later, the President, his senior staff, and other civilian officials expressed dismay at apparent attempts to influence the military's preferred course of action, partly by making the case outside normal policy channels and partly by a failure to provide a range of feasible options.[20] Several events fueled this perception. An article from *The Washington Post* on September 4 quoted General Petraeus as saying that success in Afghanistan was unlikely without many more troops. In a presentation given in London on October 1 to the International Institute for Strategic Studies, McChrystal affirmed his recommended COIN strategy and his request for troops, publicly airing his preferred course of action and refuting others in advance of any presidential decision. More damaging, however, was the leak of McChrystal's strategic assessment to the media, which essentially predicted the war would be lost if ISAF was not heavily reinforced.[21] In his memoirs, Secretary Gates described the President as "infuriated."[22] Though neither saw any calculated plan, both Gates and Chairman Mullen expressed frustration at these media missteps.

Understanding this period requires a grasp of a number of dynamic interactions. The Obama administration was new, its national security team still shaking out. The President, vice president, chief of staff and secretary of state had just come from Congress, where aggressive questioning in committee was the norm, a sharp contrast to the previous eight years. As most new administrations are, the Obama team was keen to

20 "During September several events fractured what little trust remained between the senior military and the president and his staff." Gates, p. 367.

21 See Bob Woodward, "McChrystal: 'More forces or mission failure,'" *The Washington Post*, September 21, 2009.

22 Gates, p. 368. The chapter dealing with these events is called "Afghanistan: A house divided."

assert "civilian control." In contrast, the secretary of defense, CJCS, and CENTCOM commander had long experience, their views shaped by years of involvement in the Long War and particularly by the perceived success of the surge in Iraq. Though a new four-star, General McChrystal had served extensively in both Iraq and Afghanistan as the JSOC commander and probably felt he had been given a mandate to move in a new direction as McKiernan's replacement. These and other factors contributed to quite different frames of reference and, at times, a clash of perspectives that proved difficult for all concerned.[23]

The final decision, to add an additional 30,000 troops to ISAF to resource a population-centric COIN strategy, was announced by the President at West Point on December 1. With NATO force additions, the total, surged coalition force was 140,000 personnel. This gave McChrystal much of what he had asked for, albeit with a limited timeline; the surge troops would redeploy in only eighteen months. However, the bruising contest had lingering effects. When a *Rolling Stone* article quoting McChrystal aides as critical and even contemptuous of White House officials was published six months later, McChrystal was relieved and retired, as McKiernan had been, barely a year into his tour. At least in part, the President's decision had its roots in the civil-military conflict of the previous fall.[24]

As with the invasion of Iraq in 2003 and the Iraq surge in 2006, these events represent policy and strategy-making and civil-military relations at their most complex and challenging. We ascribe no unworthy motives to any of the key players. What seems to be clear, however, is that a perception formed in the minds of senior White House staff that the military had failed to bring forward realistic and feasible options, limiting serious consideration to only one, and that it had attempted to influence the outcome by trying the case in the media, circumventing the normal policy process. These unfortunate developments affected both policy and

23 Interview with Ambassador Doug Lute, March 10, 2015.

24 Woodward, *Obama's wars* (New York: Simon and Schuster, 2010), pp. 372–373.

strategy and fed lingering resentments that would prove deleterious in the months and years to come.

In considering from a strategic perspective the key lessons from the Long War, the scholar is almost compelled to say something about America's long history with counterinsurgency. Its roots in the American experience are deep. Where successful, as in the settling of the west and in the Philippines, the methods used were often brutal and indiscriminate. More recently, in Vietnam, Iraq, and Afghanistan, our experiences have, on the whole, been difficult, costly, and indecisive. The ability of the enemy to fight from sanctuary, his unwillingness to present himself for destruction by our superior technology, the incapacity of host governments, and the loss of public support occasioned by protracted and indecisive combat all militated against clear-cut success. The historical record of large-scale, foreign expeditionary forces in counterinsurgencies is a poor one. While small scale advise and assist missions have often been successful, large-scale, expeditionary force counter-insurgency efforts do not play to American strengths and, if experience is any guide, are not likely to lead to success in securing U.S. strategic objectives.

While this essay attempts an assessment of American involvement in Iraq and Afghanistan from 2001 to the present, it will be years before a full accounting is possible. The foregoing discussion, nevertheless, sheds light on U.S. successes and failures and suggests the following concluding thoughts for consideration.

Military participation in national decision-making is necessary and essential but inherently difficult and friction-prone. Part of this derives from normal civil-military tension, but many instances in the Long War show unnecessary misunderstandings. Civilian national security decision-makers can benefit from a better understanding of the complexity of military strategy and the military's need for planning guidance. Senior military officers for their part require a deep understanding of the policy/interagency process, an appreciation for the perspectives and frames of reference of civilian counterparts, and a willingness to embrace and not resist the complexities and challenges inherent in our system of civilian control.

Vigorous debate and a clear presentation of military perspectives is essential for informed and successful strategy. Best military advice should be provided, nested within a larger appreciation of the strategic context and its political, economic, diplomatic and informational dimensions. This conversation must be carried on in confidence, respecting the prerogatives of civilian leaders with whom the ultimate decision rests.

In most cases, civilian leaders will look for a range of feasible options from the military, framed by clear cost and risk estimates, each of which can achieve the policy objective. In cases where the objective is unclear or unachievable, military leaders should press for clarity or state clearly that available resources cannot support a successful outcome. In so doing, it is helpful to consider that, in general, civilian policymakers do not come from a military planning background and that formulating specific goals and objectives is often an iterative process based on discussion and consensus.

In crafting policy and strategy, sound, well-considered ideas matter and can often carry the day. Though time is a scarce and precious resource, prior preparation and rehearsal is always a good investment. Informed and articulate advocacy has a quality all its own, and skilled communicators with a persuasive message are more likely to win acceptance.

Policy and strategy takes place in an operating universe that is highly sensitive to budget, election, and news cycles. Career military officers are not always attuned to these realities, where civilian policymakers are. Awareness and flexibility with respect to this reality improves the quality and utility of military advice.

The art of generalship at the highest levels must encompass an ability to understand and adapt to different presidential and secretarial leadership styles and modes. Within a general interagency framework, each constructs decision settings composed of personalities and processes they find most helpful and congenial. These may, and often will, vary significantly from one administration to the next. At the four-star level, the ability to comprehend and adapt to different civilian leadership styles is critical and may spell the difference between success and failure.

Four-star generals and admirals are virtually by definition masters of service and joint warfighting, but, at the most senior levels, other attributes are necessary as well, such as interagency acumen; media savvy; a detailed understanding of congressional relations and the defense planning, programming, and budgeting system; and skill in multinational environments. Normal career development patterns do not always provide opportunities to build these competencies. In a number of the examples discussed in this volume, gaps in these skill sets contributed to poor outcomes which might have been prevented either by different professional development and military and civilian education opportunities or by applying more refined selection criteria for specific, very high-level positions.

At its core, policy and strategy are all about making very hard decisions, potentially raising issues of great moral or ethical significance. Admirals and generals do not, of course, set aside personal and professional core values when they reach the pinnacle of responsibility. A strong moral compass is imperative when considering questions of war and peace. There is no settled body of thought on how senior military officers should act when faced with decisions by higher authority that, while technically legal, may violate deeply-held moral or ethical convictions. This will be a highly personal decision. Should a senior military leader face this dilemma, an honest and straightforward discussion must take place. Should the conflict be unresolvable, the officer may disassociate themselves from further involvement by requesting reassignment or retirement.

National security decision making is a highly personal endeavor relying heavily on trust relationships. These may take years to build but can be lost overnight. In this regard, General Powell's admonition—"never let your ego get so close to your position that when your position goes, your ego goes with it"—is useful. The interagency at its apex is no place for hot tempers or the easily annoyed. A calm and steady temperament can be a real advantage. Today's policy adversary may be tomorrow's policy ally. As much as possible, senior leaders will find it advantageous to maintain good working relationships with civilian partners, even when—or perhaps especially when—they find themselves on opposite sides of the issue.

If Afghanistan and Iraq are any guide, future wars will present the national security decision-maker with problems that will challenge their minds and their souls. A lesson here for future senior officers is that there is no substitute for life-long learning. The study of history, a broad grasp of all the instruments of national power with their strengths and weaknesses, confidence and a decisive character, and a fair portion of prudence and humility are all helpful when dealing with future commitments and challenges. There are no easy days for four-stars and few simple problems. Ultimately, they must deal with life and death decisions on a big stage. And, while history does not repeat itself, there are age-old patterns that senior officers and politicians must always face. Let us hope that the lessons we encountered may become lessons learned.

14 NATO in Crisis

With its twenty-nine member states, 900 million people, and $1 trillion in defense spending, the North Atlantic Treaty Organization is by far the largest, oldest and most capable defensive alliance in the world. For almost seventy years, NATO has been at the center of US national security and an indispensable component of international peace, playing an enormous role during the Cold War and beyond. Today, NATO faces enormous pressure from an aggressive Russia, massive immigration, wavering Allies—and an irresolute and critical American president. What can be done?[1]

For some six decades, the North Atlantic Treaty Organization (NATO) has been the backbone of American national security, second only to nuclear deterrence as a guarantor of peace in Europe and a major force for global stability. Today, the Alliance is threatened in ways seldom before seen, and its survival as Europe's "principal security provider" is at growing risk. NATO, beset within and without, is in trouble.

The most obvious threat is Russia, an aggressive and revanchist power intent on reasserting control and influence over its "near abroad" and regaining its place as a great power on the world stage.[2] In the past decade, Russia has retooled its military and professionalized its officer corps, now

1 The author is indebted to Francesca Buratti of the NATO Defense College for her invaluable assistance in preparing this chapter.

2 "Facing Russia's strategic challenge: Security developments from the Baltic Sea to the Black Sea," 2017, p. 6, *European Parliament Directorate General for External Policies* Issue Paper.

seasoned by years of combat in Chechnya, Georgia, the Donbas, and Syria. Repeatedly, Russia has defied international conventions and used force to change international boundaries.[3] As the primary energy provider for much of eastern and central Europe, Russia can use energy security as a tool and weapon to coerce Europe, a trend that will only accelerate with the completion of the Nordstream II project.[4]

Though not ten feet tall, Russia is formidable militarily, especially along its periphery.[5] In addition to modern ships, tanks and planes, newer Russian systems like the S-400 air defense system, the Kalibr anti-ship missile, and the Iskander surface-to-surface missile system are among the best in the world. Regular large-scale exercises keep Russia's ability to mass large forces quickly honed. The recently reconstituted 1st Guards Tank Army in the Western Military District is an offensive formation with great mobility and striking power, posing a dangerous threat to NATO's thinly guarded eastern flank. In the High North, an expanded Russian military presence and aggressive diplomacy has raised tensions.[6] Meanwhile, Russian air and missile defenses based in the Kaliningrad exclave and on the Black Sea present a stout anti-access/area denial capability that confounds NATO planners.[7]

Hard Russian military power is complemented by a sophisticated capability to conduct disinformation, propaganda, subversion, and cyber operations. With deep roots in Russian and Soviet history, now modernized and hi-tech, this ability to operate in the "gray zone" just below the kinetic level is deeply destabilizing. Throughout NATO, Russia finances far right candidates and parties, penetrates allied intelligence services, conducts targeted assassinations, and mounts effective information

3 Douglas E. Schoen and Evan Roth Smith (2016), *Putin's master plan*, p. vii, New York, NY: Encounter Books.

4 Gabriel Collins, (2017, July 18), "Russia's use of the energy weapon in Europe," *Baker Institute for Public 1/15/2019Policy Issue Brief*, Rice University.

5 Dimitri Trenin (2016, May/June), "The revival of the Russian military," *Foreign Affairs*.

6 James Stavridis (2018, May 4), "Avoiding a Cold War in the high north," *Bloomberg Opinion*.

7 The Russian Federation withdrew from the Treaty of Conventional Armed Forces in Europe in March 2015. "Facing Russia's Strategic Challenge," p. 11.

campaigns that sow doubt and dissension among Allies and partners. The intent is to foster instability and loss of confidence in democratic institutions and to drive wedges between the United States and Europe, and between European states themselves.[8] The scale and reach of Russian information operations is impressive, reaching even into the 2016 American elections with dismaying effects.[9]

What accounts for Russia's aggressive and confrontational behavior? The short answer is that Vladimir Putin derives real political benefits from confrontation with the West. His narrative, often citing "broken promises," holds that Russia has been disrespected, encircled by an encroaching NATO, and denied its rightful place as an international power of the first rank.[10] More than anything, Putin fears color revolutions on his doorstep that might threaten his rule; prospering, Western-oriented democracies on former Soviet soil must be suppressed, as in Ukraine.[11] While Western sanctions have bitten deeply, Putin's military adventurism, ruthlessness and apparent successes have helped him consolidate control of the state. He is thus incentivized to confront rather than conciliate.

Inside the Alliance, NATO faces huge challenges, its ability to weather a resurgent Russian-challenge hindered by massive refugee flows, a spillover from the Syrian civil war that has profoundly affected European polities. In Germany, Angela Merkel's attempts to cope with these waves of immigration resulted in severe punishment at the polls and the weakest Merkel governing coalition ever.[12] In Italy, fierce opposition to immigration propelled the anti-establishment, Eurosceptic Five Star movement and rightest, populist Lega—both reportedly supported by Moscow—to power in 2018. Across Europe, the rise of right-wing parties

8 Bruce McClintock and Andrew Radin (2017, May 5), "Russia in action, short of war," *US News & World Report*.

9 Julia Ioffe (2018, January/February), "What Putin really wants," *The Atlantic Monthly*.

10 Fyodor Lukyanov (2016, May/June), "Putin's foreign policy," *Foreign Affairs*.

11 Fiona Hill and Steven Pifer (2016, October 6), "Dealing with a simmering Ukraine-Russia conflict," *The Brookings Institution*.

12 Douglas Murray (2018, July 12), "Borderline disorder," *The National Review*.

is encouraged by Russia, damaging NATO and EU cohesion.[13] Along the southern flank, Greece, Italy, and Spain are more concerned with refugees and fragile, leveraged economies than with Russia.

In Hungary, the Czech Republic, Bulgaria, and Romania, democratic backsliding and corruption have reached alarming proportions, aided by Russian subversion.[14] Poland is staunchly anti-Russian but also under scrutiny for alleged suppression of journalists and a politicized judiciary. The Nordics are stable and prosperous, but their tiny militaries and proximity to Russia militate against strong measures. Across Europe, a primary objective for Putin is collapsing the sanctions regime put in place following Russia's occupation of Crimea and intervention in Ukraine. From all appearances, this movement is gaining traction.[15]

In concert with these dynamics, Turkey's slide towards authoritarianism is introducing further division inside NATO. With no real prospect of EU membership in the offing, Turkey's relations with its European neighbors (and particularly Germany, with its large Turkish minority) have badly deteriorated. U.S.-Turkish relations are also as parlous as they have ever been. Erdogan's belief in U.S. complicity in the 2016 coup; American harboring of Fethullah Gulen; support for the Kurdistan Workers' Party (PKK) as proxies in the Counter-ISIL campaign; and congressional opposition to Turkey's pending acquisition of the Russian S-400 (which will trigger automatic sanctions) have crippled the bilateral relationship. Much of Turkey's diplomatic angst and military capacity is directed against NATO ally Greece, a far weaker neighbor which poses no threat. Inside NATO structures, Turkey is chronically difficult, leveraging its status as a NATO member to play out its disputes over Cyprus. With its traditionally

13 William A. Galston (2018, March 8), "The rise of European populism and the collapse of the center-left," *The Brookings Institution.*

14 "In Europe, as in most other parts of the world, democracy is retreating and autocracy is gaining." Staffan I. Lindberg (2018, July 24), "The nature of democratic backsliding in Europe," *Carnegie Europe.*

15 Bulgaria, Cyprus, the Czech Republic, Greece, Italy, Slovenia, and Hungary support lifting Russian sanctions. "Russian sanctions and EU member states," *Association of Accredited Public Policy Advocates to the European Union*, November 9, 2017.

secular military neutered, many now doubt that in a confrontation with Russia, Turkey would fight in the Baltic states or contest Russia in the Black Sea. With little prospect for improved relations, Turkey may now pose the worst kind of insider threat to NATO: an unreliable, angry ally more likely to disrupt the Alliance than to support it.[16]

In the midst of these cares, President Trump's election and the intense focus on burden sharing that followed has badly shaken NATO. To be fair, the issue predates the election. American presidents for decades have bemoaned European free riding, and Secretary of Defense Bob Gates's famous June 2011 scolding at a NATO defense ministerial was a strong shot across the bow. Nevertheless, the pounding that President Trump has delivered on burden sharing is unprecedented. In public and private, he has pressured the Allies on their lack of progress towards reaching the "2% of GDP by 2024" defense spending goals agreed to at the 2014 Wales Summit. Claiming repeatedly that the U.S. covers "90%" of NATO's costs, President Trump has at times threatened to revisit U.S. troop presence in Europe should the performance of the Allies not improve, calling into question American commitment to Article 5 of the Washington Treaty ("an attack on one is an attack on all").[17]

On paper, their response has not been impressive. Of America's twenty-eight NATO Allies, only seven met the 2% goal as of 2018, and only fifteen have published nationally approved plans to get there.[18] Germany, at only 1.2% and with no apparent intent to grow beyond 1.5%, is a special target. No single issue has created deeper rifts between the U.S. and its European Allies than this.[19]

On closer examination, however, a different picture emerges. At the

16 Austin Bay (2018, July 31), "Erdogan's Turkey and NATO," *The Hoover Institution, 52.*

17 Eileen Sullivan (2018, July), "Trump questions the core of NATO," *The New York Times.*

18 See "Information on defence expenditures," *North Atlantic Treaty Organization*, July 10, 2018.

19 "The question of where the Western defense pact fits into a 21st century in which Europeans disagree among themselves, as well as with the United States, on economic, trade and immigration issues, and in which the world is undergoing a basic realignment with the rise of Asia, has led some to consider a new arrangement." Missy Ryan and Greg Jaffe (2018, June 29), "US Assessing the Cost of Keeping Troops in Germany," *The Washington Post.*

height of the Cold War, West Germany fielded twelve combat divisions, more than the entire U.S. Army today. Defense spending by the Big Three (Germany, France and the UK) remained well above 3% of GDP through the 1980s (for the UK, peaking at 11% in the 1950s). NATO defense spending today, excluding the United States, is some four times higher than Russia's.[20] Should it reach the 2% goal, Germany alone would outspend Russia on defense. Our Allies currently field twice as many tanks, warplanes, and warships as Russia, while their troops outnumber Russia's by more than half a million. Far from paying 90% of NATO's costs, the U.S. contributes only 22% of NATO's operating budget. Even if the entire U.S. defense budget was placed in the NATO account—an absurd notion, as most American defense spending is focused outside of Europe—the US share of overall NATO defense spending would only amount to 70%.

Our Allies are well aware of these facts, which leaves them wondering what America's real game is. While the 2% goal receives most of the press play, the real danger lies not in how much the Allies spend on defense, but how well or poorly they spend it. No issue is more worrisome than NATO readiness. NATO fields dozens of large headquarters with hundreds of generals, but few combat ready formations. Even the largest and best European Allies—the French, the British, and the Germans—would require months to put a single combat division in the field. (Poland, a bright spot in an otherwise depressing picture, can put ten times as many tanks in the field as Germany, with only one third the defense budget.) Prosperous Allies like Norway, Belgium, and Denmark field armies that contain, in their entirety, perhaps a single regular brigade. Greece, a proud member of the 2% club, spends most of its defense budget on salaries and pensions. Across the Alliance, funding for training, maintenance, and ammunition is well short of requirements. Interoperability, despite years

20 Total Alliance spending on defense for 2018 excluding the United States was $262B. "Information on defence expenditures." According to the Stockholm International Peace Research Institute, the Russian defense budget for 2017 was estimated at $61B.

of emphasis, is still a work in progress.[21]

In short, readiness—not burden sharing—is the more pressing concern. Low readiness undermines deterrence, a worry complicated by the fact that NATO is poorly postured to defend along its eastern boundary, closest to Russia. The Baltic states, for example, are defended by small local defense forces and a few "tripwire" NATO battalions. A single U.S. rotational heavy brigade is posted in Poland, backed up by one Stryker brigade in Germany and a small airborne brigade based in Italy, supported by only six fighter squadrons. Long range fires and theater air defense are modest at best. This small footprint, mostly positioned hundreds of kilometers from the border with Russia, can do little to combat Russian aggression on NATO's eastern flank in the early days and weeks of a potential conflict. In fact it would take months to marshal a force strong enough to defend the Baltic states. By then, their fall and occupation would be a *fait accompli.*[22]

For much of the Cold War, NATO relied upon the U.S. nuclear deterrent to offset its conventional inferiority. Many continue to believe that the American nuclear umbrella will ensure deterrence. Yet much has changed in the last three decades. While U.S. strategic nuclear forces remain powerful and survivable, American tactical nuclear forces have been gutted since 1990. All land and sea based tactical nuclear systems have been eliminated, leaving only a modest capability to deliver tactical nuclear weapons by air. (The 2018 Nuclear Posture Review does contemplate reviving sea-based nuclear cruise missiles in the future) In contrast, Russia maintains a powerful tactical nuclear capability that can be delivered from a variety of air, sea and ground-based platforms. Its nuclear forces are rehearsed and ready, backed up by an intimidating "Escalate to Deescalate" nuclear doctrine.[23] In a limited conventional incursion on

21 See Franklin D. Kramer and Hans Binnendijk (2018, February), "Meeting the Russian Conventional Challenge," *The Atlantic Council.*

22 David A. Shlapak and Michael W. Johnson (2016), "Reinforcing deterrence on NATO's eastern flank: Wargaming the defense of the Baltics," The Rand Corporation.

23 "Russia's national security policies, strategy, and doctrine include an emphasis on the threat

NATO territory, it is doubtful that Putin fears a nuclear response. In any case the Russian Federation possesses flexible nuclear options below the strategic threshold that give it clear advantages in this scenario.

Since President Trump's inauguration, these issues have played out against a hectic backdrop of disputes and friction. Harsh criticism of the European Union (twenty-two European countries are in both NATO and the EU); the U.S. departure from the Paris Climate Accords and the Iran Nuclear Deal; U.S. opposition to the Russian-German Nordstream II pipeline project; pending withdrawal from the INF Treaty; and serious trade disputes with key Allies have all combined to threaten and degrade NATO unity and cohesion.[24] Individually, these policy decisions may or may not have merit, though cabinet officers are known to have recommended against many of them. Collectively, they have dismayed our closest and strongest Allies and caused them profound political distress. (Deteriorating relations with the EU and with France have prompted President Macron to openly criticize NATO and call for an independent European security architecture, opening serious fissures in the Alliance.) NATO Secretary General Jens Stoltenberg has performed well in attempting to calm these roiling waters, but he is rowing upstream. In the face of multiple shocks, what could we expect from NATO in the event of a serious crisis?

A pressing worry must be the likelihood of achieving consensus. NATO cannot act without unanimity, and an expanded NATO with twenty-nine members is far more unwieldy than NATO at twelve or sixteen. For example, in the event of a Russian foray against the Baltics, with their sizeable Russian populations, it may not be realistic to assume that all twenty-nine Allies will vote for war with Russia, a war that could cost tens of billions, cause hundreds of thousands of casualties and could—if NATO forces were successful—push Russia towards nuclear escalation. Even in Washington, U.S. leaders might consider other less costly options

of limited nuclear escalation . . . Moscow threatens and exercises limited nuclear first use [to] end a conflict on terms favorable to Russia." *Nuclear Posture Review 2018*, p. 30, Department of Defense.

24 James M. Goldgeier (2018, July 10), "Trump goes to Europe," *Council on Foreign Relations.*

to avoid general war, at the cost of ceding back a few of the smallest, newest member states.

Of course, many believe that NATO's potential capabilities, conventional and nuclear, are enough to deter Russian aggression. The foregoing discussion suggests that there are sound reasons to be concerned. Still, why would Putin risk war with NATO?

For an answer, it may be useful to consider deterrence theory. The essence of effective deterrence is to threaten costs that outweigh the likely benefits of a contemplated course of action. Here, the prize for a limited, local incursion on NATO territory is enormous—nothing less than the fracturing of the Alliance altogether. If Putin reckons that that the Alliance cannot defend its Baltic members at the outset, will not use nuclear weapons, and will not vote to marshal the forces required to take them back, then NATO's deterrence regime is in tatters.[25]

There are other reasons for Putin to covet the Baltic states. They represent exactly the kind of prosperous Western democracies emerging on former Soviet territory he is known to detest and fear. They each possess ethnic Russian minorities, especially Estonia and Latvia. They stand between Mother Russia and Kaliningrad, home of the Russian Baltic fleet, but isolated and separated by 300 km from the Russian border. As Newt Gingrich declaimed during the 2016 campaign, Estonia is virtually a "suburb" of St. Petersburg, Russia's western capital. Wrenching the Baltic states from NATO control would restore the strategic depth Russia lost in the 1990s and sound the death knell for future NATO expansion for Ukraine, Georgia, and Macedonia. These are tempting rewards.[26]

This thesis is supported by the fact that on numerous occasions, Putin

25 According to Lieutenant General Ben Hodges, Commander of U.S. Army Europe until 2018, " the Kremlin wants a seat at the high table and they do that by undermining the alliance . . . the way that they do that is show that the alliance cannot protect one of its members, show that we are too slow, that we can't deter that sort of attack. If you accept that premise, then they might do a limited attack to demonstrate that NATO cannot protect its members." Patrick Tucker (2018, July 8), "This is how Russia could test NATO," *Defense One*.

26 See Ulrich Kuhn, "Russian interests and strategy," in *Preventing Escalation in the Baltics* (2018, March 28), Carnegie Endowment for International Peace.

has used military force successfully. In each case, the U.S. and NATO responded with rhetoric, sanctions and symbolic military assistance but nothing resembling concerted, tangible action. In the cases of Georgia, Crimea, the Donbas, and Syria the use of force (from the Russian perspective) has been successful without eliciting an excessively painful response from the West. Indeed, Putin is well on his way to pocketing these gains while forcing a return to "normal" diplomatic relations.

In every generation NATO must prove, to its populations as well as to its adversaries, that it remains both relevant and capable. Interludes of peace always bring forth calls for reallocation of defense resources for domestic priorities. Thankfully, the bitter national and world wars fought on European soil in the last century have receded. But generations of peace have also taught many Europeans to abhor the thought of military service. Too often, European militaries and their societies do not expect to have to actually fight. Here there is danger and vulnerability.

What then can be done? As the leader of the Alliance, the U.S. can do a better job of managing the relationship. We gain little by hectoring Allies to do things that would drive some of their leaders from office—an unreasonable expectation. We can work harder to encourage Allies to spend their money more usefully to promote actual capability and interoperability, turning "smart defense" from rhetoric to reality. We can strengthen our tactical nuclear capabilities to enhance deterrence and provide options below the strategic nuclear threshold. We can field stronger forces on the ground in Europe and place them closer to threatened areas. We can better organize ourselves, and our Allies, to combat Russian malign influence, an area where we are woefully inadequate. And we can more thoughtfully consult and consider how individual policy decisions may impact Alliance unity and cohesion—the true center of gravity for NATO.

For their part, our Allies should understand that fairer burden sharing commands widespread support with the American electorate. They must show clear progress on this front. Readiness and capability shortfalls must be solved urgently if NATO is to be more than a talking shop. NATO's cumbersome decision making processes should be streamlined and reformed. On NATO operations like Afghanistan, NATO billets must be filled to requirements. With a GDP equal to the U.S., and twice the population, Europe must step up in its own defense if the Alliance is to survive and prosper.

An immediate need is to strengthen NATO deterrence by positioning stronger forces in or near our most threatened Allies.[27] In 2017, Poland extended a generous offer to host a U.S. heavy division at its expense on Polish territory (Germany charges about $5B per year for U.S. use of its facilities).[28] While hardly posing an offensive threat, this force would go far to strengthen conventional deterrence on the eastern flank, dramatically shortening response times and altering the overwhelming conventional superiority now enjoyed by the Russian military. These moves should be accompanied by locating prepositioned stocks, not in Germany and the Netherlands (far from the point of need), but in western Poland beyond the range of most Russian rocket and missile systems. As before with Exercise Reforger, NATO should annually exercise its ability to reinforce the eastern flank. A key part of this effort must be to improve military mobility, now obstructed by infrastructure shortfalls and bureaucratic impediments.

To date, most of these moves have been resisted by U.S. officials as "provocative" or "risky." Many Allies argue the same. To be sure, Moscow protests if a single American soldier sets foot anywhere on the territory of the former U.S.S.R. or Warsaw Pact. It is in Russia's interest to keep the newer NATO members as weakly defended as possible. But simple prudence suggests that what most provokes Putin is weakness, not strength. Here the first order challenge is not how to prevail in a military

27 See "Providing for the common defense: Assessment and recommendations of the National Defense Strategy Commission," p. ix, Washington, D.C.: GPO.

28 See *Proposal for a US permanent presence in Poland 2018*, Ministry of National Defense, Republic of Poland.

conflict with Russia. There will be no real winners in a showdown with a great nuclear power. The real challenge is how to ensure that conflict never happens through a deterrence regime worthy of the name. That rests on concrete power and credibility. On this front, there is much work to be done.

The good news is that NATO holds most of the cards. Its combined GDP is some twenty times greater than Russia's, and its overall defense spending some fourteen times greater. NATO's sixty Allies and official partners constitute about 75% of the military capacity on the planet, and their combined populations dwarf Russia's. In key capabilities like theater logistics, strategic airlift, ISR, and sealift, NATO swamps Russia's modest holdings, while the list of Russia's allies is both short and unimpressive.[29]

Nevertheless, NATO must have the will to compete, and the US must lead and encourage, not berate and disrupt. The unity of the Alliance is at stake. And it is worth remembering that America's support for NATO is not based on any altruism. It is decisively in our national interest to combine and cooperate with like-minded and wealthy Allies who share our values and strategic interests. This strategic calculus saw us through perhaps the most dangerous period in world history. It can surely steer us through our present difficulties.

29 See *The military balance 2018*, International Institute for Strategic Studies.

About the Author

R. D. Hooker, Jr., is a University Professor and the Theodore Roosevelt Chair in National Security Affairs at the National Defense University. During his career, he served three tours as a director and senior director with the National Security Council; as director for the Institute for National Strategic Studies at NDU; and as dean of the NATO Defense College in Rome. A former White House Fellow, he taught at West Point and at the National War College in Washington, D.C. Colonel Hooker graduated from the U.S. Military Academy in 1981 and holds a Ph.D. in international relations from the University of Virginia. He served for 30 years in the United States Army as an infantry officer and participated in combat operations in Grenada, Somalia, Kosovo, Iraq, and Afghanistan, including command of a parachute brigade in Baghdad.

Contributors

Joseph J. Collins is a professor at the National Defense University. A retired Army colonel, he is a former deputy assistant secretary of Defense for Stability Operations and holds a doctorate in international relations from Columbia University. Colonel Collins served extensively in the offices of the Chief of Staff of the Army and the Chairman of the Joint Chiefs and is the author of *Understanding War in Afghanistan.*

H. R. McMaster served as national security adviser from 2017–2018, culminating a distinguished military career as a combat leader and soldier/statesman. Lieutenant General McMaster was awarded the Silver Star, Bronze Star with V, and the Purple Heart in a career that saw service in the Gulf War, Afghanistan, and Iraq, where he commanded the 3rd Armored Cavalry Regiment. A West Point graduate, he holds a doctorate in history from the University of North Carolina and is the author of the award-winning *Dereliction of Duty.*

Ricky L. Waddell is a lieutenant general in the U.S. Army Reserve and assistant to the Chairman of the Joint Chiefs. He served as deputy national security adviser from 2017-2018 following extensive service in Iraq and Afghanistan. A West Point graduate and Rhodes Scholar, he holds a doctorate in international affairs from Columbia University, taught at West Point, and commanded the 76th Division. He is the author of *Wars Then & Now.*

Index

A

Abgal sub-clan, 218n4, 222n21
Abkhazia, 26n44, 76
Able Archer exercise, 96
Acheson, Dean, 63
Adams, J.M.G.M., 208
Addis Ababa, 224n25
Aden, 85
ad hoc groups, 63, 67
Adolphus, Gustavus, 84
aerospace domain. *see* airpower
Afghanistan. *see also* Operation Enduring Freedom; special operations forces (SOF); U.S. Marine Corps
  Afghan National Security Forces (ANSF), 51
  airpower and, XIXn11, 162–171
  American withdrawal from, 30
  Bush (George W.) administration and, 69, 264n17
  civilians embedded in, 37
  Cold War and, 80
  future of, 256–259
  lessons of, 107, 253–255, 262, 268, 271
  low-tech opponents in, 76, 186–188
  military force in, XVIII, 82
  NATO and, 282

Obama administration and, 42n4, 192, 265–267
post-9/11, XIV–XVI, 18–20
as simultaneous conflict, 194
Soviet invasion of, 94n8, 130
surge in, 45–46
theatre joint force commanders, 9
understanding war and, 74, 86
U.S. interventions in, 33n66–34, 51
Afghan National Security Forces (ANSF), 51
Africa
Bush (George W.) administration and, 69
ethnic feuds in, 89
failed states in, 21
growth of war in, 77, 256
Horn of Africa, 220
post-9/11, XIV
as regional priority, 38
Somalia, 218
in WWII, 11
Aidid, Mohamed Farah
defeat of Barre by, 219
Mahdi as enemy of, 222n21
militia of, 233–234
Oakley and, 221, 223n21
political and strategic objectives, 236–237
U.S. Forces and, 50n13, 222, 225–231
airpower. *see also* strategic bombing
in Afghanistan and Iraq, 162, 186–187
AirSea Battle, 244–245
American edge in capabilities, 177–178
civilian casualties (CIVCAS) and, 170

description of, 146n2–147
deterrence and, 101
in Gulf War, 106, 159–160
in Korean conflict, 151
in Kosovo, 161
limitations of, 49
national defense and, 180–182
naval, 143
theorists of, 158
AirSea Battle, 243–247, 250, 252
Alaska Purchase, 6
al-Askari mosque, 263
Albania, 18, 162
Albright, Madeleine, 223, 224
Al Firdos command post, 159
Algeria, 85
Allen, Richard, 203–204
Allied Forces. *see* Allies
Allies. *see also individual countries*
Churchill on, 53
defense spending of, 276–277
deterrence and, 281–282
NATO, 274, 279
Warsaw Pact and, 142
WWII, 11
al-Qaeda
affiliates of, 255–257
definition of war and, 76
lack of statehood, 6
low tech capabilities of, XXII, 166
military operations against, 86

U.S. response, 163–164
al-Qaeda in Iraq (AQI), 255
Alsace-Lorraine, 80
American Civil War, 6–7, 83, 112, 121
American Exceptionalism, 5
American Revolution, 4–5, 111, 119
American Revolutionary War, 6
"American Way of War," 34, 45
Angola, 94n8, 130, 205
anti-access/area denial (A2/AD)
China and, 248
debate about, 241
defenses, 179
Russian, 273
U.S. policy, 246
Antigua, 208n34
antitank mines, 227
Arabian Gulf, 37
Arab-Israeli relations, 21, 106
Arab Socialist Ba'ath Party, 87, 260
Arab Spring, XV, 53
Arctic Circle, 135n19
Ardennes, 53
Argentina, 102
Armenia, 26n44
Armistice, 9
Army of the Potomac, 121n29
Asia, 10, 18, 37–38. *see also individual countries*
Aspin, Les, 71n23, 224n23, 231
Assistant to the President for National Security Affairs. *see* National Security Adviser (NSA)

Association of Southeast Asian Nations (ASEAN), 241
asymmetric threats, 22–24. *see also* terrorism
Augustine, 82
Austin, Hudson, 206
Australia, 178n87, 221n11, 249
Azerbaijan, 26n44

B

B36 controversy, 119
Baath Party. *see* Arab Socialist Ba'ath Party
Bacevich, Andrew, 109n2, 118
backbenchers (in interagency), 62
"bad wars," 35, 45
Balkans, 18, 28, 89, 217
Balkan Wars, XIV
ballistic missiles
  AirSea Battle and, 243–244
  defending against, 98
  Gulf War, 244–245
  ICBM launchers, 143, 174n71
  intercontinental ballistic missiles, 156
  Jupiter, 156
  in Maritime Strategy, 139
  qualitative gap of adversaries in, 85
  of regional powers, 196
  of Russia, 273
  Scud missiles in Gulf War as, 159–160n41
  Soviet, 93, 96, 133, 135n17–136, 142
  Strategic Air Command (SAC), 156
  U.S.S.R. threat of, 93, 96
Baltic states

NATO and, 130, 276
Russia as threat to, 26–27, 37, 278–280
Barbados, 206–210
Barents Sea, 133–134, 139, 144
Barre, Siad, 218–219, 222n20
Battle of the Black Sea, 233, 239
Battle of Kosovo Polje, 80
Battle of Lepanto, 82
Battle of Midway, 150n12
Bay of Pigs, 213
Bear Islands, 135
Beirut, 33n66–34, 212. *see also* Lebanon
Belgium, 178n87, 221n11, 277
Belize, 208n34
bellicosity, 9, 124
Berlin Wall, 16
Betros, Lance, 116
Biden Jr., Joseph Robinette ("Joe"), 70
the Big Three, 277
bipolarity
demise of, 89, 92, 104
establishment of, 14
multi-polarity and, 21
Bir, Cevik, 255
Bish, Milan, 207–208
Bishop, Maurice, 202, 205–206, 211
Black Hawk Down (Mark Bowden), 231
MH60 Black Hawks, 218n3, 227–228, 232–234, 239n69
Black Sea, 273, 276
Boot, Max, 34n68
Borneo, 248

Bosnia
persistence of war and, 73
Somalia and, 217
stabilization deployments in, XIV, XXI, 18, 33n66–34, 194n31
U.S. troop levels and, 118
Botswana, 221n11
Bowden, Mark, 231
Briley, Donovan, 235n55
Britain. *see* United Kingdom
Brodie, Bernard, 155
Budapest Memorandum (1994), 26
Budget Control Act (2011), 20, 175, 192
budgets, 9, 42, 197
Bulgaria, 202n2, 275
burden sharing, 276, 278, 281
Burke, Arleigh, 126
Bush, George H.W.
Deputies Committee and, 65n16
Grenada planning by, 208
military service of, 120n26
NSC of, 58–59
Panama invasion, XXI
Somalia and, XIV, 217, 220, 223–224
Bush, George W.
Iraq invasion and, XVIn5
Iraq surge decision of, 51–52, 262, 263–265
military capabilities required by, 184
military service of, 120n26
NSC of, 69, 71n24
two-war construct and, 184
Butler, William O., 121n29–122

C

Cambodia, 153
Canada, 6, 178n87, 221n11
Caribbean Commonwealth, 208n34
Caribbean Development Bank, 205
Caribbean Sea, 211. *see also individual countries*
Carter, Jimmy
  Carter Doctrine, 105
  foreign policy of, 205
  Iran hostage rescue, 215
  military service of, 120n26–120n27
  NSC of, 57
  Somalia and, 231
  Strategic Air Command and, 157
Center for Strategic and Budgetary Assessments (CSBA), 243–244. *see also* AirSea Battle
Central Asia, 77, 89
Central Intelligence Agency (CIA), 204, 209–210, 213–214. *see also* interagency
Chairman of the Joint Chiefs of Staff (CJCS). *see also* Joint Chiefs of Staff
  authorities of, XVII
  in Clinton administration, 118
  considerations of, 261
  determining policy outcomes, 127
  Grenada and, 206–207, 211–212
  on Iraq and Afghanistan, 187
  National Military Strategy, 41, 130
  participation urged by JFK, 124
  Powell as, XX, 33n67, 191n22
  recommendation in Somalia, 231
  responsibilities of, 263

Vessey as, 212
Vietnam conflict and, 44n7
Charles, Eugenia, 209
Chechnya, 73, 76, 85n15, 273
Cheney, Richard Bruce ("Dick"), 70–71n24
China
AD/A2 and, 179
agendas of, 25
airpower of, 161, 178n85–n186, 188n15
AirSea Battle, 243–247
communist movements supported by, 85
Communist Party and, 250
insurgencies and, 85n15
in Korean War, 151–152
military size of, 189
modern warfare of, 84–85
nationalists, 13
national security and, XXII
nuclear weapons of, 98
Offshore Control and, 247–252
Rebalance to Asia and, 23
as revanchist, XV, XV–XVI
rise of, 241–243
as rising power, 89, 196
Second Rocket Artillery Corps, 244
Vietnam War advisers of, 155
in WWII, 9–10
Churchill, Winston, 40, 41n2, 81
civilian casualties (CIVCAS)
in Afghanistan, 170, 171
impatience with, 79

in Iraq, 165
in Korea, 151
in Kosovo, 161
societal views on, 181
in Vietnam, 155
in WWII, 149
civil-military relations
in Bush (George W.) administration, 265
in Clinton administration, 118
defined, 110–110n3
history of politicians from military, 119–125
importance of, 268
modern, 37–38, 127–128
noncommissioned officers (NCOs), 113
in Obama administration, 266–267
party affiliation and, 114–117
policy and strategy, 268–271
politico-military interface and, 260–262
Somalia and, 221
strategy and, 31n58, 43–44, 51–53
views of, 109, 111–115
in wars, 48–49, 54
clans (in Somalia), 76
Clark, Mark, 152
Clark, William P., 203–204, 206n24
Clark Air Base, 32
Clausewitz, Carol von
character of war and, 83
defined strategy as, 41
long war and, 258
nature of war and, 75n5--78

persistence of war and, 91
rationality of, 86–87
statesman and commander, XXIIIn16, 45, 77, 261
tendency of war and, 47–48
trinity of, 80–81
Western views of war and, 73–74
Clinton, William J. ("Bill")
civil-military relations and, 117n22
fallout from Somalia, XIV, 50n13, 217, 226–227, 236–237
military under, 115, 184
NSC of, 61, 69–70
policy in Somalia, 223–224
CNN, 79, 219. *see also* 24-hour news cycle
coalition warfare, XX, 11, 87–88
Coard, Bernard, 206, 226
Cohen, Eliot, 109n2
Cold War
airpower in, 155–157
American success in, XIII–XIV
bipolarity and, 103–104
civil-military relations and, 114
conflicts since, 89
deterrence in, 95–96
non-aligned states in, 14
post-, decline in forces, 30, 191
post-, increase in conflict, XIV, 17
Soviet interventions, 23, 218
strategy changes since, 29–30, 32–33
weapons of, 174n71
color revolutions, 274
Combined Air Operations Center (CAOC), 168

commander-in-chief, XX, 7, 118, 264
The Command of the Air (1921), 147
Congo, 73
Congress
  aviation procurement and, 180
  defense mobilization and, 194
  Grenada and, 212
  House Armed Services Committee, 206
  military appointments and, 125
  military policy and, 127
  NSA and, 61
  popular support in Somalia and, 237, 240
  threats to U.S. and, 33
Constantine the Great, 83
Constitution of the United States, 56
contractors (civilian), 180, 184, 191, 254
conventional deterrence, 94–95, 103–108, 282. *see also* deterrence; nuclear deterrence
counterinsurgency (COIN). *see also* insurgency
  in Afghanistan, 257–258
  airpower and, 166–167, 170
  American history of, XVI, XXI, 268
  Army and, 183
  complications of, 168
  efficacy of, 263–267
  post-9/11, XIV, 19–20, 253–255
countermine, 245
counter-proliferation, 31
Counter-Terrorism, XIV, 19, 265
Court of St. James, 117n22
Creech Air Force Base, 167

Crete, 11
Crimea, 26–27, 275, 281
Crisis Pre-Planning Group, 204, 208
Cromwell, 84
Crowe, William, 117n22, 120n27
cruise missiles. *see also* ballistic missiles; missiles
American dominance of, 188n15
defending against, 134
Ground Launched Cruise Missiles (GLCMs), 96
in Gulf War, 160
in Maritime Strategy, 139
in Nuclear Posture Review (2018), 278
strategic application of, 136
on submarines, 143
Tomahawk, 141
Crusader artillery system, 127n40
Cuba
Grenada and, 202n3, 212–213
Marxist insurgencies and, 204–205
Rough Riders and, 120n26
Spain and, 6
Cuban Missile Crisis
as alarming, 96, 202
bipolarity and, 15
importance of alternate opinions and, 52
Kennedy administration and, 215
cyber attacks
AirSea Battle and, 245, 251
civil-military relations and, 37–38
as emerging threat, XV, 20
Russian capabilities, 273–274

security environment and, 34–35
U.S. Navy on, 144
as weapon of mass destruction, 22–23
Cyprus, 275
Czech Republic, 275

D

Darood clan, 218n4
Dawkins, Pete, 120n27
Dayton Accords, XIV, 18
defense budgets
aircraft and, 175, 178–180, 195
AirSea Battle and, 244
herculean task of managing, 49
military personnel, 192–193
national security definition and, 21
NATO and, 277
pressure on, 184
reductions of, 17, 195
Strategic Air Command (SAC), 155–156
U.S. investments and, XXII
of U.S. Navy, 195n33, 244–245
defense industry, 192
Defense Intelligence Agency (DIA), 256
Defense Strategic Guidance, 4
de-housing (bombing), 150
democracy, 25, 47–48, 90–91
Democracy, Human Rights, and International Organizations (DNSA for), 69
Democratic Party
historical lessons and, 34

military size and, 184, 188n16
NSC in administrations of, 70
partisan politics and, 50
U.S. Armed Forces and, 116–117
Democratic Republic of Congo, 76
Democratic Republic of Vietnam, 153–155
Dempsey, Martin, XVIIn6
Denmark, 138n31, 178n87, 277
Department of Defense (DoD)
Department of Defense (DoD), 41
financial obligations of, 255n4
National Command Authority, 194
Deptula, David, 157
Deputies Committees (DC), 57, 59–65, 67–68
Deputy Assistant Secretary of Defense for the Middle East, 65
Deputy National Security Adviser (DNSA), 61, 69. *see also* National Security Adviser (NSA)
Deputy Secretary of Defense, 65
Desert One, XXI
desiderata, 34
deterrence. *see also* conventional deterrence; nuclear deterrence
Army's capstone manual on, 101n14
definition of, 92
deterrence theory, 105, 280
in Gulf War, 105–107
studies on, 95n10
Dewey, Thomas E., 120, 122n33
Diego Garcia, 32n59
Direct Air Support Center, 154
Dominica, 208n34–209
Dominican Republic, XXI, 34, 194n31

Donbas, 273, 281
Donilon, Tom, 242
Douhet, Giulio
affirmed in wars, 172
Kosovo and, 161
theories of, 146–148n7, 150
U.S. Air Force and, XVIII
the draft, 112. *see also* selective service
drawdown of 1990s, 17–18, 112, 114, 118
drones, 166–167n58
Dulles, John Foster, 93

E

East China Sea, 23
East Germany, 202n2
economy, 11–12, 28–29n52, 78, 251–253
Eisenhower, Dwight D.
Cold War Era forces of, 93
directive to, 51
NSC staff of, 68
presidency and, 120, 122n33
electoral politics, 122
El Salvador, 85, 205, 213
England. *see* United Kingdom
Enlightenment, 5
entitlement spending, XXII
Erdogan, Recep Tayyip, 275
Estimated Global Nuclear Warhead Inventories, 2019, 97f
Estonia, 27, 280
Ethiopia, 94n8, 218, 234
Euphrates River, XIXn11

Europe. *see also individual countries*
Cold War and, 95, 156
deterrence and, 104
Organization for Security and Cooperation in Europe (OSCE), 26
as regional priority, 37–38
technology and, 84
U.S. forces in, 28
European Economic Community, 205
European Union, 23, 274–275, 279
Eurosceptic Five Star movement, 274
Executive Branch, XXII, 4, 31, 71, 122, 126–127
Exercise Reforger, 282

F

Falklands War, 102
Fallujah, 171n66
Feaver, Peter, 109n2
Fehrenbach, T. R., 75
Fethullah Gulen, Muhammed, 275
financial crisis of 2008, 23, 29
Finland, 178n87
first principles, 31, 74
First World War. *see* World War I (WWI)
Fisher, John Arbuthnot, 47
"Five Rings" theory, 165
Flexible Response, 94–95. *see also* conventional deterrence
Flynn, Michael, 35n73, 256
Fogelman, Ronald, 158
Foggy Bottom, 59
Ford, Gerald R., 57, 120n26
Forrest, Nathan Bedford, 75

France
airpower of, 178n87
Cold War defense spending and, 277
Indochina and, 88, 100
NATO and, 85
size of armed forces, XVII
Somalia and, 221n11
wars and, 80
in WWII, 9–10
Frank, Ray, 235n55
Frederick the Great, 84
Fremont, John C., 120–122
French Revolution, 5
French Second Empire, 80
fusionist view, 124

G

Ganjgal Valley (2009), 169
Garfield, James A., 120
Garrison, William F., 227–228, 232–233, 235, 237–238
Gates, Robert
on F-35 program, 174
Iraq troop surge and, 187n12, 264–265n19
NATO scolding by, 276
in NSC, 71n23
in Obama administration, 266
spending by European allies and, 30
George, Alexander, 214
Georgia (country), 26, 273, 280–281
Germany
airpower of (NATO), 178n87

Cold War defense spending and, 277
creation of, 80
forces in Afghanistan of, 170
modern warfare of, 84
NATO and, 274–278, 282
quality of army, 11n16
size of armed forces, XVII
submarines and, 135n19
U.S. grand strategy, 8–10
Vietnam War bombing and, 155
WWII and, 12–13n20, 23, 51
WWII bombing of, 149–150, 155
Ghali, Boutros-Boutros, 222n20
Ghani administration, 255
Gingrich, Newt, 280
GIUK Gap, 132f–133
Giumale, Sharif Hassan, 234
Global War on Terrorism, 82
Glorious Revolution (1688), 80
"good wars," 35, 45
grand strategy
after 9/11, 18
China and, 23–25
and definitions, 41
at end of WWII, 12–13
framework of, 37–39
fundamentals of, 20–21
NATO and, 26–28
soft power and, 31
unilateral action and, 33
in WWI, 9

in WWII, 11
Grant, Ulysses S., 119–120n28
Gray, Colin, 42
gray zones, XVn4, 273
Great War. *see* World War I (WWI)
Greece, 10, 275, 277
Greenland, 132f–133
Greenland-Iceland-Norway Gap, 132f–135, 141
Grenada
intelligence on, 205–208
interagency and, 212–214
lessons from, 215–216
U.S. intervention in, XXI, 33n66–34, 51, 194n31, 201–202, 209
Grotius, 82
Guam, 8
Guantanamo Bay, 213
Gulf War. *see also* Iraq; Iraq War (2003-2011); Operation Desert Storm
airpower and, 157–160
American success in, XIII, XVI
deterrence and, 93, 105–107
Land Component Command and, XIXn11
refugee assistance following, 18
support for, 33–34, 46
U.S. land power, 189–190
U.S. success in, XIX–XX, 51

H

Habr Gidr sub-clan, 218n4, 221n14–222n21, 225
Hagee, Mike, 221
Hagel, Chuck, XVIIn6, 71n23, 185n7
Haig, Alexander, 71n24, 204n14

Haiti, 18, 28, 34, 194n31
Hamas, 76
Hancock, Winfield S., 119–120
hard power, 30–31
Harrison, William Henry, 120
Hawiye clan, 218n4
Hays, Rutherford B., 120
Heraclitus, XXIV
Hezbollah, 76
Hoar, Joseph, 231
homeland defense, 37
Horner, Charles, 159
Howe, Jonathan, 224–226
human rights, 25
Human Rights Watch, 161n44, 165
Hungary, 178n87, 275
Huntington, Samuel P., 34n69, 111n12, 119, 124
Hussein, Saddam
    as counterweight to Iran, 257
    deterrence and, 106–107
    economic disruption of, 23
    Kuwait invasion and, 53
    Operation Iraqi Freedom, 165
    success against, XV, 255
    U.S. objectives and, 86
    WMDs and, 260

I

Iceland, 130, 132f–135, 138n31
idealist theory of politics, 124
improvised explosive device (IED), 166, 227

India
in Allied air forces, 178–178n87
grand strategy and, 24n39
India-Pakistan War of 1971, 104n19
potential for conflict and, 243
as rising power, 21
Indian Wars, 6
India-Pakistan War of 1971, 104n19
Indochina, 80, 100
Indonesia, 24n39, 178n87
industrial revolution, 83
INF Treaty, 279
insurgency. *see also* counterinsurgency (COIN)
2014 Afghanistan mission statement, 51
as asymmetric threat, 22, 37
in El Salvador, 205
Islamist, 258
in post-Saddam Iraq, 260
protracted, 85
intelligence, surveillance, and reconnaissance (ISR)
airpower and, 148, 160
Coalition, 164
latest-generation, 166
NATO capacity, 283
U.S. lead in air power and, 178
interagency. *see also* National Security Adviser (NSA); National Security Council (NSC)
Afghanistan and, 259, 265
airpower and, 148
civil-military relations, 46, 48, 128, 268–271
frustrations with, 66–68

Grenada and, 206–207, 214n49–216
NSA and, 58, 60–61
NSC and, 57–59, 62–63, 71
in Obama administration, 64–65
in Regan administration, 204
roles played in, 213–214
Somalia and, 224
U.S. safety and, 71–72
wars and, 48–49
Western Hemisphere actions of, 215
working groups of, 59
Inter-American Affairs, 206
intercontinental ballistic missiles. *see* ballistic missiles
International Institute for Strategic Studies, 266
International Monetary Fund, 205
Intifada, 106
Iran
hostage rescue mission (1980), 215, 238n63
interagency working groups and, 64
Iraq and orbit of, 255, 257
military size of, 189
national security and, XXII
as politically fragile, 89
as rogue state, XV
in Syria, 25
U.S. military operations and, 87
Iran-Contra, 57, 66n18
Iran Nuclear Deal, 279
Iraq. *see also* Gulf War; Iraq War (2003-2011); Operation Desert Storm
American withdrawal from, 28, 30
counterinsurgency (COIN), XIV–XVI

influenced by Iran, 257
interagency working groups and, 64
Iraq Steering Group, 62
Iraq Study Group, 263
ISIS and, 25
lack of statehood and, 76
lessons of, 107, 262
low-tech opponents in, 186
ongoing military operations in, 74, 82, 86–87
theatre joint force commanders, 9
U.S. military in, XIXn11, 19–20, 51
U.S. misunderstanding of, 46
Iraq Steering Group, 61–62
Iraq War (2003-2011). *see also* Iraq; Operation Desert Storm; Operation Enduring Freedom
airpower in Iraq and Afghanistan and, 162–168
assumptions about, 52
civil-military relations, 265, 267–268
cost of, 54
interagency and, 65
Multinational Forces Iraq (MNF-I), 188n14
public support and, 33n66–34
surge, 45, 262
U.S. forces in, 189
Isaaq clan, 218n4
Iskander surface to surface missile system, 273
Islam, 89–90, 233, 256, 258. *see also* Shia Islam; Sunni Islam; Sunni-Shia divide
Islamic Republic of Afghanistan. *see* Afghanistan
Islamic State in Iraq and Syria (ISIS), 25–26, 256
Islamic State in Iraq and the Levant (ISIL), 64, 255, 257, 275

isolationism, 13, 36–37
Israel
  Arab-Israeli relations, 21, 106
  Gulf War and, 160
  land power of, 190
  modern warfare of, 84
Italy
  airpower of (NATO), 178n87
  Cold War and, 156
  NATO and, 274–275, 278
  security arrangements with, 32n59
  Somalia and, 221n11, 234
  WWII and, 9, 149
Izzo, Gerry, 234n52

J

Jackson, Andrew, 120–121
Jackson, Michael, 162
Jamaica, 208n34
Jan Mayen, 135
Japan
  airpower of (as ally), 178n87
  Department of Defense (DoD), 32n59
  Imperial Japanese Navy, 12, 150
  islands of, 248
  moving closer to U.S., 241
  Offshore Control and, 248
  Sea of Japan, 84
  strategic bombing of, 150, 152, 155
  as U.S. ally, 243
  U.S. Navy based in, 30n53

U.S. security shield, 105n20
Vietnam War bombing and, 155
Western postwar order and, 28n49
WWII and, 9–10, 12–13, 23
jihad/jihadists, 87
Johnson, Louis, 71n23, 119
Johnson, Lyndon B., 94, 118, 120n26, 153, 194
Johnston, Robert, 220–221
Joint Chiefs of Staff. *see also* Chairman of the Joint Chiefs of Staff (CJCS)
civil-military relations, 124, 261
in Clinton administration, 118
Grenada, 206–207, 211
Iraq and, 263
operational control (OPCON), 211
Somalia, 231
Vietnam War and, 44n7
Joint Force Air Component Commander (JFACC), 168
Joint Operational Access Concept (JOAC), 246
Joint Special Operations Command (JSOC), 227–228, 238n63. *see also* special operations forces (SOF)
Joint Special Operations TF (JSOTF), 227
Jordan, 167

K

Kabul. *see* Afghanistan
Kaliningrad, 274, 280
Kandahar, 163
Kashmir, 77
Keating (2009), 169
Keegan, John, 74, 161
Kennan, George, 36

Kennedy, John F. (JFK)
Cuban Missile Crisis and, 215
on military in policy making, 124, 261
military service of, 120n26
nuclear deterrence and, 93–94
Kennedy, Robert, 52
Kenya, 219
Key West Naval Station, 205n21
Kidd, Admiral Isaac, 143
Kissinger, Henry, 38
Kittani, Ismat, 222n20
Kohn, Richard, 109n2, 119
Kola Peninsula, 136–137, 139–140
Konduz, Afghanistan, 170
Korean War
airpower in, 151–156, 177
American narratives and, 6
deterrence and, 92–93
as "hot war," 155–156
as peace enforcement action, 32n59–33n66, 34
as protracted military operation, 51, 80
threats from, 196
U.S. and, XIX, 13–15, 189–190
Kosovo War
Kosovo Liberation Army, 161
Military Technical Agreement, 18
NATO airpower and, 161–162
NATO and, 52
persistence of war and, 73
Somalia and, 217
stabilization deployments in, XIV, XXI, 18, 194n31

weak support for intervention in, 33n66–34
Kurdistan, 26
Kurdistan Worker's Party (PKK), 275
Kuwait
Gulf War and, 93, 158–160
invaded by Iraq, 105–107
security arrangements with, 32n59
world order and, 23

L

Laden, Osama bin, 163
Lake, Anthony, 71n23, 224n23
Land Component Command, XIXn11, 194
land power
air power and, 148, 172
American history of, 34
assessment of, 183
importance of, 193–198
in Iraq and Afghanistan, 186–189
National Security Strategy, 183
Operation Desert Storm and, 160
sea power and, 145
Laos, 153, 154
Latin America. *see* Cuba; Cuban Missile Crisis; Grenada
Latvia, 26, 280
Lebanon. *see also* Beirut
interventions in, XXI, 15, 33n66
Lebanon War (1982), 190
Somalia and, 208, 230
Legislative Branch, 31, 122, 126–127
LeMay, Curtis, 120n27, 126

Liberia, 18, 238n63
Libya
NATO and, 172
Obama and, 30, 50
personnel in Grenada from, 202n2
U.S. interventions and, 24, 28
Liddell Hart, B. H., 41
Lincoln, Abraham, XXIV, 50, 121n29–122n31
lines of effort (LoE), 64
Lithuania, 26
MH6 Little Bird helicopters, 227–228
AH6 Little Bird helicopters, 227–228, 234n51
Lockheed, 174n72–175
Logan, John, 122n31
Lombok, 249
Long Range Strike Bomber (LRS-B), 179
Louisiana Purchase, 6
Low Countries, 130
Luzon, 249

M

MacArthur, Douglas, 12, 119–120, 122n33, 125
Macedonia, 18, 280
MacFarlane, Robert C. (Bud), 203–204, 207n30, 209–210
Mahan, Alfred Thayer, 102
Mahdi, Ali, 221n14–223n21
Majertain sub-clan, 218n4
Malacca dilemma, 249, 252
Malay Peninsula, 248
Malaysia, 24n39, 85, 178n87, 226
Maliki, Nouri al-, 262–263

maritime power. *see* Maritime Strategy (1980s); U.S. Navy
Maritime Strategy (1980s)
approach, 129–131
final phase of, 136–139
Phase Two of, 133
requirements of, 141–145
Marshall, George, 119–120n27
Mayaguez incident, 24
McChrystal, Stanley, 170, 265–267
McClellan, George B., 119–112
McClernand, John A., 122
McKiernan, David, 120, 265, 267
McKinley, William, 120n26
McMaster, H. R., 44n7, 69
McNamara, Robert, 71n23, 118
Mead, Dave, 229–230, 238
media leaks, 42
Medicins sans Frontiers, 228
Mediterranean, 138n31
Medvedev, Dmitry, 32n60
Meese, Ed, 204n12
Melians, 81
Mellinthin, F. W. von, 80n10
Menges, Constantine, 206n24–207
Mentemeyer, Rich, 221
Merkel, Angela, 274
Mexican-American War, 6, 121
Middle East
ethnic feuds in, 89
geostrategic effects of Iraq intervention, 257
multi-polarity and, 21

as regional priority, 37–38
threats from, 196
U.S. forces in, 29
war flourishing in, 78
military
agenda of services of, 53–54
in American society, 113n13
appointments of, 124–126
conventional forces devalued and, 93
cost of war and, 54
culture in Armed Forces, 262
inter-service success in, 34
National Security Strategy, 145
political office and, 120n27
professionalism, 7, 123–124, 259
Republican Party affiliations and, 115–117
size decreasing, 184
Somalia and, 218
U.S. power and, 31–32
Military Assistance Command Vietnam (MACV), 154
military-industrial complex, 122
military leaders, 44–45. *see also* civil-military relations
Milosevic, Slobodan, 161–162
missiles. *see* ballistic missiles; cruise missiles
modern war, 145
Mogadishu. *see* Somalia
Moltke, Helmuth von, 41, 259
Mombassa, 219
Momyer, William M., 154
Mongols, 83
Monroe Doctrine, 5

Montgomery, Thomas, 225–228, 230–233, 237n61–238
Montserrat, 208n34
"morale bombing," 150
Morocco, 221n11
Moscow. *see* Russia; Soviet Union; Union of Soviet Socialist Republics
Motley, Langhorne A., 206–207
Mozambique, 94n8
mujahideen, 80
Mullen, Mike, 187n12
Murtha, John, 232n44
Murusade sub-clan, 218n4
mutually assured destruction, 22, 98
Myers, Richard B., 260n9

N

Namibia, 94n8
Napoleon, 5, 46, 80, 84
narco-trafficking, 20, 77
national debt, 255
National Defense Strategy (NDS), XXII
National Military Strategy, 4
national security. *see also* National Security Adviser (NSA); National Security Council (NSC)
  of American people, 72
  Americans on role of, 17
  civilian control and, 110
  definition of, 37
  military and, 124–125, 128
  nuclear deterrence and, 98
  threats, 20
National Security Act, 57n2, 183n2

National Security Adviser (NSA). *see also* National Security Council (NSC)
Grenada and, 202–204, 215
interagency functioning and, 68
as presidential adviser, 58–61
rebalance to Asia, 242
in Regan administration, 209
Scowcroft as, XX
Somalia and, 224
of vice president, 70
National Security Council (NSC). *see also* interagency; National Security Adviser (NSA)
in Bush (George W.) administration, 264n17
Chief of Staff, 60–61
Grenada and, 206–207, 209–210, 215
growth over time of, 66–72
Iraq and, 263
Office for Iraq and Afghanistan, 69, 264n17
in Regan administration, 203–204
Somalia and, 221
staff of, 56–58, 61–64
National Security Decision Directive (NSDD), 209
National Socialism, 80
Naval War College, 7
Near East and North Africa offices of DNSA, 69
Netherlands, 4, 138, 178n87, 282
New Jewel Movement, 205–206
New Model Army, 84
24-hour news cycle, 35, 58, 66. *see also* CNN
Nicaragua, 85, 205
Nigeria, 219

9/11
Army reorganized following, XIX
costs of wars following, 186, 255
Homeland Security Council, 69
military recruiting, 115
U.S. grand strategy and, 18–19
WMDs and, 23
Nixon, Richard, 57, 120n26, 153
non-state actors, 35, 37–38
Nordstream II pipeline project, 273, 279
Normandy, 277
North Africa, 77
North Atlantic, 134
North Atlantic Council, 26
North Atlantic Treaty Organization (NATO)
Afghanistan troop surge and, 267
airpower of, 178
Alliance with, 24, 32, 85
Allied Command Europe Mobile Force, 138
American national security and, 36
Baltic states, 26–27
burden sharing, 275–278, 280–283
challenges to, 85
in Cold War, 16
deterrence and, 243
Kosovo and, 52, 161–162
Libya and, 172
in Maritime Strategy, 134–136, 139
nuclear deterrence and, 95
Rebalance to Asia and, 30
Russia as threat, 272–274, 279–281

in strategic environment, 196
submarines, 142
Taliban and, 265
U.S. commitment, 32, 185
U.S.S.R. and, 14, 130
Washington Treaty, 25–27
Northern Alliance, 163
North Korea. *see also* Korean War; South Korea
deterrence and, 98
Grenada and, 202n2–202n3
invasion of South Korea, 13–15, 53
Korean War and, 151–154
military size of, 189
post-9/11, 19
as threat to West, XV–XVI, XXII, 89, 196
North Vietnam, 84, 153–154. *see also* Vietnam; Vietnam War
Norway
airpower of, 178n87
Allied performance in, 10
Maritime Strategy and, 130, 132–135, 138n31
Norwegian Sea, 130–131, 133–134n9, 142–144
nuclear arms race, 96
nuclear deterrence. *see also* conventional deterrence; deterrence; mutually assured destruction
arms control agreements, 32n60, 96, 104
as foundation of national security, 107–108
National Security Strategy, 98
NATO and, 283
threat of force and, 103
Nuclear Posture Review, 2018, 278
nuclear weapons. *see also* Cold War; conventional deterrence; nuclear

deterrence
arms control agreement and, 104–105
arms race and, 96
deterrence effect of, 85
end of WWII and, 12
Maritime Strategy and, 136, 141–142
NATO and, 278–280
Offshore Control and, 250
post-WWII proliferation of, 14–15
as Russian deterrent, 93–95
strategy and, 179
technology and, 135n16
threats of, 196
U.S. arsenal, 13

O

Oakley, Robert, 221–222
Obama, Barack
Afghanistan surge, 265–268
arms control agreement and, 32n60
deterrence and, 184
Libyan stance of, 30
NSC of, 61, 68–69
Rebalance to Asia, 242
U.S. Pacific allies and, 251–252
Ocean Venture (1982), 205
Office of the Vice President (OVP), 70
Offshore Control, 247–252
Ogabeni sub-clan, 218n4
O'Hanlon, Michael, 185
Oklahoma City bombing, 23n37

Operation Allied Force, 161
Operation Anaconda, 163
Operation Continue Hope, 223. *see also* United Nations Operation in Somalia II (UNOSOM II)
Operation Desert Storm. *see also* Gulf War
airpower and, 157–158, 161n44, 168, 244
as strategic success, 45, 160, 184
Operation Enduring Freedom, 163, 259
Operation Iraqi Freedom, 164–165, 259
Operation Linebacker I, 153
Operation Linebacker II, 153
Operation Northern Watch, 165
Operation Proud Deep, 153
Operation Provide Relief, 219
Operation Restore Hope, 222–223n21, 236
Operation Rolling Thunder, 153
Operation Saturate, 151
Operation Southern Watch, 165
Operation Strangle, 151
Operation Urgent Fury, 201, 212, 216
organizational thinking, XVIII, 56, 68
Organization for Security and Cooperation in Europe (OSCE), 26
Organization of Eastern Caribbean States (OECS), 208–209
organized crime, 16, 20, 22
Otto, Osman, 228

P

Pacific Ocean
airpower in WWII and, 149
American interests in, 9–10
China and, 241–242, 251

deployments in, 134
U.S. interests in, 8–12, 37, 89, 178, 194
Pacific Rim, 89
Pakistan
al-Qaeda and, 163, 167, 255–257
Somalia and, 219, 221, 226, 229–230, 234n51–236, 238n64
strategy and, 52
Palestinians, 85n15
Panama
U.S. success in, XXI, 18, 44n8, 51
U.S. withdrawal from, 28
weak support for intervention in, 33n66–34
Panama Canal, 249
Panetta, Leon, XVIIn6, 71n23
Paris Climate Accords, 279
Paris Peace Accords, 153
Pearl Harbor, 11, 40, 53
Pence, Michael Richard, 70
Pentagon
AirSea Battle and, 246
budgets, 19n28, 23n37
interagency and, 65
NSAs and, 59
Somalia and, 231
People's Republic of China. *see* China
Pershing, John J., 119–121n28
Persian Gulf, 23, 105–107, 160, 186
Petraeus, David, 255, 264, 266
Philippines
counterinsurgency and, XIV, 76, 268
as U.S. ally, 178, 241, 248

in U.S. history, 8, 32
WWII and, 12
Poindexter, John, 204
Point Salines, 213
Poland, 178n87, 275, 277–278, 282
politics, 123–127. *see also* civil-military relations; *individual civilian leaders*
Portugal, 178n87
Powell, Colin
Base Force recommendation, 118, 191
as Chairman of the Joint Chiefs, XX
civil-military relations and, 117, 119
NSC and, 71n23–24
Powell Doctrine, 48
as Secretary of State, 120n27
on temperament, 271
Thucydides and, 46
on use of force, 33n67
presidential special envoys, 64, 68
presidents, 31, 36–37, 125. *see also* National Security Adviser (NSA); National Security Council (NSC); *individual presidents*
the press, 48
Principals Committees (PC), 57–60, 62, 64–68
Prueher, Joseph, 120n27
Prussia, 46, 80, 84
public opinion, XXIII, 88, 100
Puerto Rico, 6, 8
Punic Wars, 83
Putin, Vladimir, 274–275, 279–282

Q

Qatar, XIXn11, 32n59, 168
Quadrennial Defense Reviews, 4, 20, 184

R

Rahanwein clan, 218n4
Reagan, Ronald
  grand strategy in Cold War and, 38
  Grenada and, 201–205, 207n30, 209–211, 215
  military expenditures of, 130
  military service of, 120n26
  nuclear deterrence and, 157
realist theory of politics, 124
Rebalance to Asia
  as counter to China, 23
  NATO and, 30
  of Obama administration, 242, 252
  other potential conflicts and, 37
religion, 78, 89
Republican Party
  Army affiliations with, 115–117
  historically, 34
  military size and, 184, 188n16
  NSC in administrations of, 70
  partisan politics and, 50
Responsibility to Protect (R2P), 24
Revolt of the Admirals, 119, 125, 156
Revolutionary Military Council (RMC), 206, 208, 212
Revolutionary War. *see* American Revolution
Roman empire, 83, 90
Romania, 275

Roosevelt, Franklin Delano, 10n14, 12
Roosevelt, Theodore, 120n26–121n28
Rough Riders, 120n26
rule of law, XXII, 5, 114, 257
Rumsfeld, Donald, 19, 44n6, 71n24, 262n13–264
Rusk, Dean, 13n21, 25n41
Russia. *see also* Russian Federation; Soviet Union; Union of Soviet Socialist Republics
agendas of, 25
airpower of, 177–178, 188n15
Baltic states and, 26–27, 37, 280
Chechnya and, 85
decline of, 104
deterrence and, XXI–XXII, 93, 179
disinformation of, 273–275
1st Guards Tank Army, 273
insurgencies backed by, 94, 234
Kalibr anti-ship missile, 273
Kosovo and, 162
military size of, 189
national security and, XIV, XXII
NATO and, 272–283
NSC and, 64, 69
nuclear options of, 97–98, 277–279
post-Cold War, 104
proxies of, 52
as resurgent, 21, 196
as revanchist, XV, 272
as rival, 80
security arrangements with, 24n39
Somalia and, 234

Syria and, 25
threats from, 196
Ukraine and, 26–27
Vietnam War and, 155
in Western Military District, 273
WWI and, 9
WWII and, 10–11, 80, 93
Russian Federation, XIV, 97, 177–178, 278–279. *see also* Russia; Soviet Union; Union of Soviet Socialist Republics
Rwanda
humanitarian aid to, 18
persistence of war and, 73
Rwandan genocide, XIV, 47, 50n, 217
Ryukyu Islands, 248

S

Sandinistas, 85n15
Saudi Arabia, 93, 105n20–106, 196
Scandinavia, 130
Schwarzkopf, Norman, XX, 159
Scoon, Paul, 208–209
Scott, Ridley, 231
Scott, Winfield, 119–121
Scowcroft, Brent, XX–XXI, 58–59, 71n23
Seaga, Edward, 208
sea lines of communication (SLOCs), 130, 133, 141–143
Sea of Japan, 84
Second World War. *see* World War II (WWII)
Secretary of Defense. *see also individual secretaries of defense*
ability to effect change of, XVII
NSC and, 57n2

Obama and, 267
Somalia and, 217
Vietnam War and, 154
selective service, 29n51
Senate. *see* Congress
Serbia, 161–162. *see also* Kosovo War
Shahi-Kot Valley, 163
Sheridan, Philip, 122
Sherman, John, 121n28
Sherman, William Tecumseh, 119–121n29
Shia Islam. *see also* Islam; Sunni Islam; Sunni-Shia divide
al-Askari mosque bombing, 263
divide with Sunni Islam, 21
misunderstanding of, 52, 260
political power of, 257
resistance of, 87, 166
Shinseki, Eric K., 190
shock and awe, 163
Shultz, George, 203, 207, 209–210
Sickles, Daniel E., 122
Sinai, 18
Singapore, 11, 24n39, 178n87
Slovenia, 275n15
Smith, Bedell, 120n27
soft power, 30, 77
soldier-statesman, 124
Somalia
Abdi house raid and, 225–228
clans, 76, 218n4, 223–224n25
fallout from intervention, XIV–XV, XXI, 217–219
humanitarian aid and, 18, 194n31, 220–222n18

lessons of, 50n13, 73, 235n57–240
Responsibility to Protect (R2P) and, 24
Russia and, 94n8
SNA militia and, 229–233
U.S. Army helicopters and, 227–228, 235
U.S. chain of command in, 238
U.S. withdrawal from, 28
weak support for intervention in, 33n66–34
Somali National Alliance (SNA), 218n4, 225–227
Somali National Army, 234
Somali National Movement, 218n4
Somali Patriotic Front, 218n4
Somoza regime, 205
South China Sea, 23
South Korea
airpower of, 178n87
bi-lateral security arrangements with, 24, 30, 32n59, 95, 185, 241
deterrence and, 98
invaded by North Korea, 13–15
post-9/11, 19
South Ossetia, 26n44
South Vietnam, 120n27
Southwest Asia, 18
South Yemen, 94n8
Soviet Union. *see also* Russia; Russian Federation; Union of Soviet Socialist Republics
aviation of, 134, 136
in Cold War, 23, 218
defenses of, 174n71
fall of, 16, 53, 89, 97
forces in WWII, 9–10, 13

Grenada and, 201–202, 211, 213
Korean War threat of, 152
Maritime Strategy and, 130, 136, 140–144
nuclear deterrence and, 97, 157
nuclear forces of, 94, 135n17, 156–157
post-WWII confrontation with, 15
Soviet ballistic missile submarine (SSBN), 133–135, 141n40–142
strategic bombing reserved for, 152
and unipolar world, XIVn1
U.S. lessons from Cold War, 243
Spain, 4, 6, 178n87, 275
Spanish American War, 8
Sparta, 81
special operations forces (SOF). *see also* Joint Special Operations Command (JSOC)
command of, 54
drones and, 167
in Grenada, 211
growth of, 9
in Gulf War, 159n40
in Iraq and Afghanistan, XVIII, 169
ISIS and, 26
limitations of, 45, 49, 186–187
post-9/11 growth, 19
in Somalia, 226–227, 232–235, 238n63
Special Situation Group (SSG), 204, 208
Spruance, Raymond, 150n12
statecraft, 77, 82
statehood, 76, 78–79
Steering Group, 64. *see also* interagency; Iraq Steering Group; National Security Council (NSC)

St. Kitts-Nevis, 208n34
St. Lucia, 208n34
Stockdale, James, 120n27
Stoltenberg, Jens, 279
St. Petersburg, 280
Strachan, Hew, 3n2
Strait of Tiran, 23
Straits of Hormuz, 186
Straits of Magellan, 249
Straits of Malacca, 23
Strategic Air Command (SAC), 154–157
strategic attack. *see also* airpower; Douhet, Giulio
    in Gulf War, 157, 165
    in U.S. Air Force thinking, 171, 179, 181
strategic bombing, 12, 161, 172, 181
strategic culture, 5, 34–35, 45
Strategic Deterrence Initiative (SDI), 96
strategic mobility, 101n15–102, 194n30
strategos, 41
strategy
    autonomy and, XVIII–XIX
    classical notions of, 87
    National Security Strategy, 4, 40–43
    nature of the conflict and, XXIII
    strategic assessments and, 52–53
St. Vincent, 208n34
Subic Bay, 32
submarines. *see also* U.S. Navy
    in Cold War, 135n19
    Maritime Strategy and, 133–135, 142
    Offshore Control and, 248

submarine warfare, 12
sub-Saharan Africa, 77
Sudan, 73, 76
Sunda Straits, 249
Sunni Association of Muslim Scholars, 263
Sunni Islam, 87, 166, 263. *see also* Islam; Shia Islam; Sunni-Shia divide
Sunni-Shia divide, 21, 257. *see also* Islam; Shia Islam; Sunni Islam
superpowers. *see also* bipolarity
rivalry of, 15–17, 92, 104, 202, 211
U.S. and, 6, 12
surface-to-air missiles, 140, 143, 153
sustainability, 102
Svalbard archipelago, 135
Sweden, 76, 178n87
Syria
interagency working groups and, 64
ISIL and, 255
resistance in, 87
Russia and, 25, 273–274, 281
Syrian civil war, XIV–XV
U.S. inaction in, 30

T

Tailhook scandal, 126n39
Taiwan, 13, 24n39, 178n87, 248
Taliban. *see also* U.S. Navy
9/11 attacks and, 163–164
Afghanistan surge and, 265
counterinsurgency of, 166
as low tech opponent, XXII
military operations against, 86, 255n4–257

persistence of, XV, 257
statehood of, 76
Tampa, XVIII, XIXn11, 168, 227, 231, 237. *see also* U.S. Central Command (CENTCOM)
Task Force (TF) Ranger, 227–228, 232–233, 235–237, 239
Taylor, Maxwell, 94, 120n27
Taylor, Wesley, 213n46
Taylor, Zachary, 119–121
technology
advantage in Somalia of, 239
airpower and, 147, 162–163, 176, 181
in Gulf War, 157
in Iraq and Afghanistan, 186
nuclear weapons and, 93–94
war and, 83–85, 88–89
terrorism
Iraq and, 25
Islamist, 256, 258–259
post-Cold War, XIV, 16
as threat to West, 20–22, 88–89
war and, 74, 76–77, 82, 85
Third World, 94, 98, 211
threshold of pain, 100
Thucydides, 46
Tibet, 78
Tora Bora, 163
Torrejon Air Base, 32
Total Obligational Authority, 193
Tower Commission, 66n18
transnational actors, 76
Transnistria, 76

Trinidad, 208n34
Trotsky, Leon, 73
Truman, Harry S., XVIIn6, XVIIn8, 119, 120n26–27, 125
Trump, Donald
  great power competition and, 184
  NATO and, 276, 279
  NSC of, 61, 69
Turkey, 64, 156, 178n87, 275n15–276
Turner, Stansfield, 120n27
24-hour news cycle, 58, 66, 270

U

Ukraine, 26, 275, 280
Unified Task Force (UNITAF), 220–221, 223n22, 225, 233n48
Union of Soviet Socialist Republics. *see also* Russia; Russian Federation; Soviet Union
  airpower in Cold War and, 155
  communist movements supported by, 15, 205
  deterrence and, 243
  Maritime Strategy and, 130
  nuclear deterrence and, 14, 93, 96
  Strategic Air Command and, 157
  territory of former, 282
  war flourishing in former, 77
  as WWII ally, 10
United Kingdom
  airpower of, 178n87
  as Ally, 9–10
  British Special Air Services, 244
  as coalition partner, 87
  Cold War defense spending and, 277

deterrence and, 102
grand strategy and, 4
Gulf War and, 244
history of, 23
in Operation Iraqi Freedom, 165
role of British Crown, 4
size of armed forces, XVII
Somalia and, 218
submarines and, 135n19
United Kingdom/Netherlands Amphibious Force, 138
WWII and, 10n14–11, 149n10–150, 155
United Nations (UN)
Aidid and, 222n20
in Cold War, 16
in Korean War, 152
National Security Council ambassador, 66
in 1990s, 17
Somalia and, 220–222n20
supporting use of force of, 103
U.S. in coalition of, 32, 196
United Nations Operation in Somalia (UNOSOM I), 219
United Nations Operation in Somalia II (UNOSOM II), 223–227, 229–230, 232, 235–236
United Somali Congress, 218n4
United States. *see also* civil-military relations
challenged by revanchist powers, XV
Communist bloc confrontation, 13
cost of war and, 85
counterinsurgency and, XVI
as dominant power, 31, 33, 186
future wars and, 88–91

geostrategic position of, 257
nuclear technology of, 96
in Republic, 3
technological sophistication of, 85
as world power, 9, 28–29
WWII and, 11–13
unity of command, XVIII, 54, 237n62
UN Security Council Resolution (UNSCR), 224
UN Supreme Commander, 152
U.S. Air Force
airpower of, 147–148n8, 188n15
AirSea Battle and, 243–244, 246
aviation in other military services and, 176–177
Chief of Staff, 158, 246
Cold War deterrence, 93
cooperation, 153, 186–187
deployments of, 86
in 1990s drawdown, 17
drones and, 167
at end of WWII, 12
Far East Air Force Bomber Command of, 151–151n16
Gulf War and, 157–161
as independent service, XVIII, 148n8, 152
in Iraq and Afghanistan, XIXn11, 164, 168–169, 171n66
in Korean War, 151
Long Range Strike Bomber (LRS-B), 179
post-9/11, 19–20, 163
program costs of, 172–176, 179, 181–182
in Somalia, 227, 232
stealth strikes and, 136n21
strategic attack doctrine, 171–172, 179, 181–182

strategic bomber force of, 154–157
strength of, 182, 195
Vietnam War and, 153–155n27
U.S. Armed Forces. *see also* U.S. Army; U.S. Marine Corps; U.S. National Guard; U.S. Navy; *individual services*
airpower lead of, 178n88
budgets, 9
challenges to, 35
civil-military relations and, 110, 128
in Gulf War, 107
mobility of, 102
national security and, 17–18
post-9/11, 255–256
strategy and, 261–262
U.S. Army
Air Corps Tactical School, 172
American Civil War, 122
budgets, XVIII–XIX, 20, 50, 192–193, 195
capabilities, 196–198
challenges, 186–188
Chief of Staff, 120, 191n22, 229
coalition forces and, 87
Command and General Staff College, 7
cooperation with U.S. Air Force, 152–154, 156, 177, 181–182
defense reductions, 20
Delta Force, 227, 232, 235
dependence of, XIX, 9
deployments, 86
deterrence and, 101n14
drawdown of, XIV, 17, 192–195
evolution of, 7

Gulf War and, XIXn11, 159–160n43
helicopters used by, 169, 177n83
history of, 7
in Iraq and Afghanistan, 186–189, 194
Iraq surge, 264
Marine Expeditionary Force and, 139n32
mission of, 183–184
NATO spending and, 277
nuclear weapons, 105
101st Air Assault Division, 163
operational control (OPCON), 152
personnel costs, 193n29
post-9/11, XIX, 18–19
professionals of, 121–122
size of, 190–196
staff, 225
Supreme Allied Commander Europe, 162
10th Mountain Division, 163, 220, 225n27, 229, 235, 237, 239
Vietnam War and, 153n22
WWII and, 9–10
U.S. Army Air Force, 9–10, 150
U.S. Army Corps of Engineers, 193
U.S. Army National Guard, 19n30, 187–188n13, 193–195n35
U.S. Army Rangers, 213, 228, 232, 235
U.S. Army's School of Application for Infantry and Cavalry, 7
U.S. Army War College, 232n44
U.S. Central Command (CENTCOM)
Air Component Commander, XIXn11
airpower commanded by, 159, 166n55, 168
in Iraq and Afghanistan, XIXn11
Iraqi invasion of Kuwait and, 105

Iraq War and, 260n9
Obama administration and, 267–268
operational control (OPCON), 237
operations in Somalia, 225, 227, 230–231
Schwarzkopf and, XX
U.S. CENTCOM Intelligence Support Element (CISE), 228
U.S. Congress
Chairman of the Joint Chiefs of Staff (CJCS), 127
Committee on the Conduct of the War, 122–124
defense industry and, 180, 192
the draft and, 194
Grenada and, 211–212
Iraq surge and, 50, 264–265
national threats and, 33
NSA and, 61
Obama administration and, 266
Somalia and, 237, 240
U.S. Department of Defense (DoD). *see also* interagency
acquisition budget of, 191–193
aircraft produced for, 174n72, 180
forces worldwide, 31n59
Grenada and, 206–207, 209, 212
industry and, 180
Iraq and Afghanistan deployments and, 188n13
National Command Authority and, 194
NSC and, 57
post-9/11, 19n29, 163
in Regan administration, 204
U.S. National Guard and, 188n13
U.S. Department of State (DOS). *see also* interagency
Grenada and, 208n35–210, 212–213

National Security Decision Directive (NSDD), 209
in Regan administration, 203–204
U.S. grand strategy. *see* grand strategy
U.S. Liaison Office (USLO), 221
U.S. Marine Corps
air forces of, XVII–XVIII
aviation of, 153n22–154, 172–174, 177, 182
Beirut terrorist attack on, 201
Budget Control Act and, 20
CAS response time in Korea of, 152n19
Combat Out Post (COP), 169
deployments of, 30n53, 187–188, 194
1st Marine Expeditionary Force, 191, 220
Grenada and, 208
Gulf War and, XIXn11, 159n40–160n43
in Iraq and Afghanistan, 87, 187n11–189, 194
Iraq surge, 264
Libya and, XXI
Marine Air Ground Task Force (MAGTF), 169
Marine Air Wing, 150, 152, 160
Marine Amphibious Force, 131
Marine Expeditionary Brigade (MEB), 132–133
Marine Expeditionary Force, 131n6, 138n29, 138n31–32
in Maritime Strategy, 131–132, 137–138, 144
post-9/11 growth of, 19
size of, 190–196
in Somalia, 220–221
strength of, XVII, 195
support troops of, 138n27
in WWII, 150
U.S. Marines. *see* U.S. Marine Corps

U.S. National Guard, 19n30, 187–188n13, 193–195n35
U.S. Navy
aircraft costs and, 173–174
aircraft of, 152–153n22, 177, 182
AirSea Battle and, 243–244
Amphibious Ready Groups, 30n53
Chief of Staff, 246
defense budget and, 195n33
Department of the Navy, 144
deployments of, 30n53, 86
in drawdown of 1990s, 17
evolution of, 7
fleet inventory of, 188n15
Gulf War and, 160n43
Maritime Strategy and, 129–130, 134, 136–137, 143
nuclear weapons, 105
post-9/11 and, 19
Pueblo incident, 24
Revolt of the Admirals and, 156
role after 9/11, 19
scaling back of, 30
sea power lead of, 32n62, 144, 188n15
strength of, XVII–XVIII, 195
submarines, 139
as supporting land power, 186–187
U.S. Striking Fleet Atlantic, 131–131n6
in WWII, 9–10, 150
U.S. Northern Command, 37
U.S. Pacific Command, 120n27, 154
U.S. Pacific Fleet, 11
U.S. Senate. *see also* U.S. Congress

NSA and, 61

V

values gap, 113–115
Versailles, 80
Vessey, John, 212
Vice Chairman of the Joint Chiefs of Staff, 65
Vice Director for Strategy, Plans, and Policy, 65
Vienna, 84
Vietnam. *see also* Vietnam War
  airpower of, 178n87
  China and, 24, 241
  disengagement from, 94
  ideology and, XV–XVI, 52
  Indochina conflicts and, 100–101
  insurgencies and, 85n15
  security arrangements with, 24n39–25n41
  U.S. misunderstanding of, 46
Vietnamization, 153
Vietnam War. *see also* Vietnam
  airpower and, 174n74, 182
  civil-military relations and, 114
  Cold War and, 15–16, 129
  deterrence and, 92, 94
  forward air controllers-airborne (FAC-As), 153–155
  insurgencies and, XXII, 84–85, 254
  political direction and, 31
  procurement costs and, 190–193
  as protracted military operation, 51, 268
  Somalia and, 230
  technology and, 157

U.S. land power, 190, 262
Vom Kriege (Carl von Clausewitz), 74, 87

W

Wales Summit (2014), 276
Wanat (2008), 169
war. *see also* Clausewitz, Carol von; *individual wars*
definition of, 73–75
nature of, 78
persistence of, 91
Powell Doctrine and, 48
unpredictability of, 81–82
war crimes, 48
Warden, John, 157–158, 164
War Department, 121
War of 1812, 5n5, 18, 121
War Powers Act, 201, 211n41
Warren, Francis E., 121n28
Warsaw Pact, 95, 142, 282
Washburne, Elihu, 121n28
Washington, George, 120
Washington Treaty, Article 5, 27, 276
Watkins, James, 120n27, 131n7
weapons of mass destruction (WMD)
definition of, 22n37
future opponents and, 88–90
Hussein and, XVIn5
non-state actors and, 20
regional powers and, 196
Saddam Hussein and, 255, 259–260
as threats to homeland, 22, 37, 38

Weigley, Russell, 109n2, 119
Weinberger, Caspar, 33n67, 71n24, 212, 214n49
Weinberger Doctrine, 48
Wesch, Michael, 109n2
the West. *see also individual countries*
    deterrence and, 100
    Islam and, 89–90
    militaries of, 84
    thoughts on war of, 82
West Germany, 277
Westmoreland, William, 120n27, 154
West Point, 267
White, Thomas, 120n27
White House Situation Room, 65, 69, 211
Wilhelmine Germany, 80
Wilson, Charles, 71n23
Wolcott, Cliff, 235n55
Wood, Leonard, 120, 122n33
Wood, Robert E., 119
World Concern, 228
World Trade Center, 23n37
World War I (WWI)
    economic integration and, 243
    support for, 33
    technology of war and, 83
    U.S. in, 9, 23
World War II (WWII)
    airpower in, 146–150
    American grand strategy, 10–11
    American strategic performance since, XVII
    Army power since, 189

Army strength at pre-WWII lows, 195
bombs used in, 154
conflicts since, XXI
conventional forces devaluation after, 93
durability of war and, 80
ground technology little advanced, 195n36–196n
historical narrative of, 40
hostile peer competitors in, 23
manufacturing output in, 29n52
origins of, 80
post-war national security concerns, 15
U.S. aircraft lost in, 149n10–n111
U.S. global alliances since, 31
U.S. military use since, 33

Y

Yemen, 76, 94n8
Yom Kippur War (1973), 104n19, 190

Z

Zaire, 221
Zarqawi, Abu Musab al-, 167

CPSIA information can be obtained
at www.ICGtesting.com
Printed in the USA
JSHW010225220720
6792JS00001B/4